SpringerBriefs in Optimization

SpringerBriefs in Optimization showcases algorithmic and theoretical techniques, case studies, and applications within the broad-based field of optimization. Manuscripts related to the ever-growing applications of optimization in applied mathematics, engineering, medicine, economics, and other applied sciences are encouraged.

More information about this series at http://www.springer.com/series/8918

Nathan Adelgren

Advancing Parametric Optimization

On Multiparametric Linear Complementarity Problems with Parameters in General Locations

Springer

Nathan Adelgren
Department of Mathematics
and Computer Science
Edinboro University
Edinboro, PA, USA

ISSN 2190-8354 ISSN 2191-575X (electronic)
SpringerBriefs in Optimization
ISBN 978-3-030-61820-9 ISBN 978-3-030-61821-6 (eBook)
https://doi.org/10.1007/978-3-030-61821-6

Mathematics Subject Classification: 90C31, 14XX

This Springer imprint is published by the registered company Springer Nature Switzerland AG
The registered company address is: Gewerbestrasse 11, 6330 Cham, Switzerland

For my girls, Ashley and Azalea

Preface

When unknown or uncertain data in an optimization problem is replaced with parameters, one obtains a multiparametric optimization problem whose optimal solution comes in the form of a function of the parameters. As the parameters' values change within a set of admissible values, the decision variable and objective function values change as well. Solving a multiparametric optimization problem involves two things: (i) finding representations of the optimal decision variables and optimal objective value as functions of the parameters and (ii) partitioning the parameter space into regions over which these function representations are valid.

Interest in parametric optimization has recently grown due to applications in optimal control and model predictive control. Another broad application of parametric optimization lies in modeling uncertainty that can affect the optimization model. Exogenous uncertainty (such as temperature, pressure, material properties in engineering, or interest rate in business applications) or feasibility uncertainty related to the fulfillment of constraints can be modeled by means of parameters that do not have a fixed nominal value but rather belong to a known, typically closed, and convex set, and the solutions to the related parametric optimization problems reveal optimal decisions for various realizations of uncertainty. Parametric optimization also offers a bridge to robust optimization, a prominent area of research, which uses smart optimization schemes to yield robust solutions arguably being preferred under the conditions of uncertainty. However, since the robust solutions correspond to specific values of uncertainty, the robust optimization approach is subsumed in parametric optimization and the latter emerges as a more universal methodology.

In this monograph, we consider the multiparametric form of the linear complementarity problem (mpLCP) in which all input data in the form of a matrix M and a vector q is permitted to be dependent on a vector of parameters taken from a bounded semi-algebraic set. The procedure proposed herein is the first we know of that is capable of solving this large class of problems, and therefore its presentation has wide reaching implications in many areas of multiparametric optimization. In particular, the proposed methodology can be utilized to solve any multiparametric linear or convex quadratic program as well as multiobjective optimization problems with any number of linear and/or convex quadratic objectives. It is also worth noting that for many of these classes of problems, specifically multiparameteric convex quadratic programs with parameters in general locations and multiobjective optimization problems with more than two convex quadratic objectives, no other solution procedure is currently known. As the applications of the aforementioned problems are extremely numerous, we believe that anyone at the graduate level or above having an interest in theoretical or applied mathematical optimization will benefit from reading this book. Additionally, the correctness of the methodology presented herein is argued using concepts from the field of Algebraic Geometry. Hence, anyone interested in the intersection of Algebraic Geometry and Operations Research will also find the contents of this book worthwhile.

Edinboro, PA, USA Nathan Adelgren

Acknowledgments

The author gratefully acknowledges the partial support by the United States Office of Naval Research through grant number N00014-16-1-2725.

The author also wishes to thank Michael Burr and Margaret M. Wiecek of Clemson University for their helpful dialogue and advice during the development of this work.

Contents

Nomenclature

θ	Parameter vector for mpLCP
$M(\theta)$	Matrix of affine functions of θ
$q(\theta)$	Vector of affine functions of θ
ϕ	Subvector of θ such that every element of ϕ is present in some element of $M(\theta)$
υ	Subvector of θ such that no element of υ is present in any element of $M(\theta)$
$M(\phi)$	Alternative notation for $M(\theta)$
$q(\phi, \upsilon)$	Alternative notation for $q(\theta)$
Θ	Parameter space
$\hat{\Theta}$	Feasible subset of parameter space
$G(\theta)$	Augmented matrix $\begin{bmatrix} I & -M(\theta) \end{bmatrix}$
$G(\phi)$	Alternative notation for $G(\theta)$
$\Theta_{\mathcal{I}}$	Set of θ in Θ for which $G(\theta)_{\bullet\mathcal{I}}$ is full column rank, given a particular index set $\mathcal{I}$
$\mathcal{B}$	Arbitrary basis—must satisfy $\lvert\mathcal{B}\rvert = \lvert q(\theta)\rvert$ and $\Theta_{\mathcal{B}} \neq \emptyset$
$\mathcal{IR}_{\mathcal{B}}$	Invariancy region associated with basis $\mathcal{B}$
$g_{\mathcal{B}}$	Positive one whenever $det\,(G(\theta)_{\bullet\mathcal{B}}) \geq 0$ for every θ in Θ; negative one otherwise
$\mathcal{C}_{\mathcal{B}}(\theta)$	Parametric complementary cone associated with basis $\mathcal{B}$
$\mathcal{C}_{\mathcal{B}}(\phi)$	Alternative notation for $\mathcal{C}_{\mathcal{B}}(\theta)$
$\mathcal{ID}_{\mathcal{B}}$	Invariant domain associated with basis $\mathcal{B}$
U	Support of $M(\theta)$
V	Complement of U
$\Phi(\upsilon)$	Set of attainable values of ϕ, given a fixed υ
$\Upsilon(\phi)$	Set of attainable values of υ, given a fixed ϕ
Φ	Set of ϕ for which there exists a υ such that (ϕ, υ) is in Θ
Υ	Set of υ for which there exists a ϕ such that (ϕ, υ) is in Θ
$\mathcal{IR}_{\mathcal{B}}(\upsilon)$	Set of ϕ for which (ϕ, υ) is in $\mathcal{IR}_{\mathcal{B}}$, given a fixed υ
$\mathcal{IR}_{\mathcal{B}}(\phi)$	Set of υ for which (ϕ, υ) is in $\mathcal{IR}_{\mathcal{B}}$, given a fixed ϕ

Ω	Arbitrary subset of Θ satisfying the Polynomial-Linear Property (Property 3.1)
$\Omega(\upsilon)$	Set of ϕ for which (ϕ, υ) is in Ω, given a fixed υ
$\Omega(\phi)$	Set of υ for which (ϕ, υ) is in Ω, given a fixed ϕ
$\Omega.\mathcal{D}$	Set of defining inequalities of Ω
$\Phi^{\Omega,\mathcal{S},\bullet}$	Set of ϕ in $\text{Proj}_U\, \Omega$ for which the set of redundant inequalities of $\Omega(\phi)$ is $\Omega.\mathcal{D} \setminus \mathcal{S}$, given a subset $\mathcal{S}$ of $\Omega.\mathcal{D}$
$\Phi^{\Omega,\bullet,d}$	Set of ϕ in $\text{Proj}_U\, \Omega$ such that the fiber of Ω resulting from fixing ϕ has dimension d
$\Phi^{\Omega,\mathcal{S},d}$	Intersection of $\Phi^{\Omega,\mathcal{S},\bullet}$ and $\Phi^{\Omega,\bullet,d}$
$\Omega^{\mathcal{S},d}$	Set of (ϕ, υ) in Ω, where ϕ is taken from $\Phi^{\Omega,\mathcal{S},d}$ and υ is taken from $\Omega(\phi)$
$\mathscr{h}^i_{\mathcal{B}}$	Hypersurface that implies the defining constraint of $\mathcal{IR}_{\mathcal{B}}$ associated with index i, i.e., the hypersurface given by $(Adj(G(\phi)_{\bullet\mathcal{B}}))_{i\bullet}\, q(\phi, \upsilon) = 0$
$\mathcal{F}_{\mathcal{B}}$	Set of indices in $\mathcal{B}$ whose associated hypersurfaces form $(k-1)$-dimensional boundaries of $\mathcal{IR}_{\mathcal{B}}$
$\mathscr{B}$	Set of feasible complementary bases that have been discovered and processed
$T_{\mathcal{B}}(\phi, \upsilon)$	Tableau associated with basis $\mathcal{B}$, i.e., the augmented matrix $\left[G(\phi)^{-1}_{\bullet\mathcal{B}} G(\phi) \mid G(\phi)^{-1}_{\bullet\mathcal{B}} q(\phi, \upsilon)\right]$
$T_{\mathcal{B}}(\phi)$	Alternate notation for left hand side entries of $T_{\mathcal{B}}(\phi, \upsilon)$
$\text{Z}_{\mathcal{B}}$	Set of indices in $\mathcal{B}$ for which the right hand side of the tableau $T_{\mathcal{B}}(\phi, \upsilon)$ is identically zero
$\text{E}_{\mathcal{B}}$	Set of indices in $\mathcal{B}$ whose associated hypersurfaces do not intersect $\mathcal{IR}_{\mathcal{B}}$
$\text{H}^i_{\mathcal{B}}$	Set of indices j in $\mathcal{B} \setminus (\text{E}_{\mathcal{B}} \cup \{i\})$ such that the intersection of $\mathscr{h}^i_{\mathcal{B}}$ and $\mathcal{IR}_{\mathcal{B}}$ is a subset of the intersection of $\mathscr{h}^j_{\mathcal{B}}$ and $\mathcal{IR}_{\mathcal{B}}$
$GCD(\mathcal{B}, i, j)$	Greatest common divisor of $(Adj(G(\phi)_{\bullet\mathcal{B}})_{i,\bullet}\, q(\phi, \upsilon)$ and $(Adj(G(\phi)_{\bullet\mathcal{B}})_{j,\bullet}\, q(\phi, \upsilon)$
$\text{D}^i_{\mathcal{B}}$	Set of j in $\mathcal{B}$ for which $GCD(\mathcal{B}, i, j)$ is a nonconstant polynomial and the intersection of $\mathscr{h}^i_{\mathcal{B}}$ and $\mathcal{IR}_{\mathcal{B}}$ is a subset of the intersection of $\mathcal{IR}_{\mathcal{B}}$ and $GCD(\mathcal{B}, i, j)$, whenever the dimension of $\mathcal{IR}_{\mathcal{B}}$ is less than k; otherwise, the set is empty
$EQ_{\mathcal{B}}$	Set of indices in $\mathcal{B}$ whose corresponding defining constraints of the phase 1 variant of $\mathcal{IR}_{\mathcal{B}}$ are binding at an optimal solution of $NLP_S(\mathcal{B})$ (5.4)

Chapter 1
Introduction

In this work we consider the multiparametric form of the Linear Complementarity Problem (LCP) in which all input data is permitted to be dependent on a vector of parameters $\theta \in \Theta \subseteq \mathbb{R}^k$. Here, Θ specifies the set of "attainable" values for θ, and we assume that it is given by a bounded semi-algebraic set (see Chapter 2, Definition 2.8). This problem is referred to as the multiparametric Linear Complementarity Problem (mpLCP). Let $\mathscr{A} = \{\alpha^\top \theta + \beta : \alpha \in \mathbb{R}^k, \beta \in \mathbb{R}\}$, the set of affine functions of θ. Then mpLCP is as follows:

> Given $M(\theta) \in \mathscr{A}^{h \times h}$, $q(\theta) \in \mathscr{A}^h$, for each $\theta \in \Theta$, find vectors $w(\theta)$ and $z(\theta)$ that satisfy the system
>
> $$\begin{aligned} w - M(\theta)z &= q(\theta) \\ w^\top z &= 0 \\ w, z &\geq 0 \end{aligned} \tag{1.1}$$
>
> or show that no such vectors exist.

mpLCP is said to be *feasible at* θ if there exist $w(\theta)$ and $z(\theta)$ that satisfy (1.1) and *infeasible at* θ otherwise. Similarly, if there exists a $\hat{\theta} \in \Theta$ at which mpLCP is feasible, then mpLCP is said to be *feasible*. mpLCP is *infeasible* if no such $\hat{\theta}$ exists. We define the set

$$\hat{\Theta} := \{\theta \in \Theta : \text{mpLCP (1.1) is feasible at } \theta\}. \tag{1.2}$$

Now, recognize that Θ must be an infinite set, otherwise mpLCP reduces to LCP. Hence, it is not possible to determine a solution to (1.1) for each $\theta \in \Theta$ individually. Instead, mpLCP is solved by partitioning the space Θ into a set of *invariancy regions* and its infeasible subset $\Theta \setminus \hat{\Theta}$. As the name "invariancy regions" suggests, within each of these regions, the representation of the solution vectors w and z as functions of θ is invariant. Note that although we refer to these sets as

N. Adelgren, *Advancing Parametric Optimization*, SpringerBriefs in Optimization,
https://doi.org/10.1007/978-3-030-61821-6_1

invariancy regions, they have also been given other names in the literature. Some examples are critical regions and validity sets. A detailed discussion on invariancy regions and their properties is provided in Chapter 2. We point out that, although the terminology is similar, the invariancy regions considered herein ought not to be confused with *invariant sets* that are employed within a variety of other fields of mathematics. Unique definitions of *invariant sets* exist for topology [54] as well as dynamical systems [46] and, by extension, model predictive control [6].

Note that the non-parametric version of LCP has the same form as (1.1) with the exception that $M(\theta)$ and $q(\theta)$ are replaced by $M \in \mathbb{R}^{h\times h}$ and $q \in \mathbb{R}^h$, respectively. LCP is a well-known problem in the literature and has been studied extensively by researchers such as Kostreva [36], Lemke [38], Murty and Yu [42], and Cottle et al. [14]. Though LCP is NP-hard in general, polynomial time algorithms exist for certain classes of the matrix M. Thus, much work has been done in order to identify various classes of matrices M that impact one's ability to solve an instance of LCP. Solution techniques for LCP are often designed for specific classes of M. For a concise list of important matrix classes, see [13]. For a detailed discussion on these classes and their impact on LCP, see [14, 42]. We will refer to many of the matrix classes discussed in these works throughout this text.

Single parametric LCP (pLCP) with a parameter present only in the $q(\theta)$ vector (i.e., $k = 1$ and $M(\theta) = M \in \mathbb{R}^{h\times h}$) was first proposed as a result of the work done by Maier [40] and has been studied extensively since. Columbano et al. [11], Gailly et al. [21], Li and Ierapetritou [39], and Adelgren and Wiecek [2] considered (1.1) with $k > 1$ and $M(\theta) = M \in \mathbb{R}^{h\times h}$. The method of Gailly et al. [21] is designed for the case in which M is a copositive-plus matrix, the methods of Columbano et al. [11] and Adelgren and Wiecek [2] are designed for instances in which M is a sufficient matrix, and the method of Li and Ierapetritou [39] works for general M. Parametric and multiparametric LCPs in which $M(\theta) \neq M \in \mathbb{R}^{h\times h}$ (i.e., the matrix $M(\theta)$ cannot be represented as a real valued matrix, as it depends on θ) have received little attention though. Interesting properties of the case in which $M(\theta) \neq M \in \mathbb{R}^{h\times h}$ were discussed by Tammer [48]. Xiao [57] and Chakraborty et al. [9] presented solution techniques for the case when $M(\theta) \neq M \in \mathbb{R}^{h\times h}$ but restricted that $M(\theta)$ be a P-matrix for all $\theta \in \Theta$. Additionally, Väliaho [51] proposed a method for solving (1.1) with $M(\theta) \neq M \in \mathbb{R}^{h\times h}$, but only for the case in which $k = 1$, and several assumptions on the structure of $M(\theta)$ and $q(\theta)$ are satisfied.

The method we propose in this work is an extension of the work done by Adelgren and Wiecek [2], who considered (1.1) for $M(\theta) = M \in \mathbb{R}^{h\times h}$ and $k > 1$. We note that while a preliminary version of this study is contained in Adelgren [1], this monograph is a result of further investigation conducted by the author. Our method solves (1.1) whenever it is feasible and the following assumptions are satisfied.

Assumption 1.1 *We assume that $M(\theta)$ is a sufficient matrix for all $\theta \in \Theta$.*

Assumption 1.2 *We assume that the interior of $\hat{\Theta}$ is connected.*

We note that in the case in which Assumption 1.2 is not satisfied, our method will identify and partition a single feasible subset of Θ that has a connected interior. For additional insight into the importance of the above assumptions, we point the reader to Section 6.1, where we present examples of instances of mpLCP that *do not* satisfy these assumptions and discuss the difficulties that this causes. Now, since we require $M(\theta)$ to be sufficient, we provide the following definition, as found in [14].

Definition 1.1 A matrix $M \in \mathbb{R}^{h\times h}$ is *column sufficient* if the following implication is satisfied:

$$(x_i(Mx)_i \leq 0 \text{ for all } i) \quad \Rightarrow \quad (x_i(Mx)_i = 0 \text{ for all } i) \tag{1.3}$$

M is said to be *row sufficient* if $M^\top$ is column sufficient. If M is both column and row sufficient, it is then called *sufficient*.

We note that although there exist finite time algorithms capable of determining whether or not a given matrix is sufficient (see, for example, [53]), in general, determining whether or not $M(\theta)$ is sufficient for all $\theta \in \Theta$ is not a trivial task. To see this, consider the following results.

Lemma 1.1 (Theorem 4.6 of Väliaho [52]) *The matrix* $A = \begin{bmatrix} 0 & B \\ C & 0 \end{bmatrix} \in \mathbb{R}^{n\times n}$, *where the zero blocks are square, is sufficient if and only if all the corresponding minors of B and $-C^\top$ have the same sign.*

Proposition 1.2 *The set of sufficient matrices is not closed under convex combinations.* □

Proof Consider the matrices $A^1 = \begin{bmatrix} 0 & 2 \\ -1 & 0 \end{bmatrix}$ and $A^2 = \begin{bmatrix} 0 & -1 \\ 2 & 0 \end{bmatrix}$. It is easy to see that both A^1 and A^2 satisfy the conditions of Lemma 1.1 and are therefore sufficient. For each $\alpha \in [0, 1]$, consider the following:

$$\alpha A^1 + (1-\alpha)A^2 = \begin{bmatrix} 0 & 2\alpha \\ -\alpha & 0 \end{bmatrix} + \begin{bmatrix} 0 & \alpha - 1 \\ 2 - 2\alpha & 0 \end{bmatrix}$$

$$= \begin{bmatrix} 0 & 3\alpha - 1 \\ 2 - 3\alpha & 0 \end{bmatrix}$$

It is clear that for all $\alpha \in (\frac{1}{3}, \frac{2}{3})$, the above matrix does not satisfy the conditions of Lemma 1.1. Thus, the above matrix is not sufficient for any $\alpha \in (\frac{1}{3}, \frac{2}{3})$. This clearly shows that the set of sufficient matrices is not closed under convex combinations.

□

The result of Proposition 1.2 also shows that the set of sufficient matrices cannot be closed under conic or linear combinations. Thus, even in the case in which Θ is translated into the nonnegative orthant and there exist sufficient matrices $M^0, M^1, \ldots, M^k \in \mathbb{R}^{h\times h}$ such that $M(\theta) = M^0 + \sum_{i=1}^{k} M^i\theta_i$, one is not guaranteed that $M(\theta)$ is sufficient for all $\theta \in \Theta$. This serves as evidence that the problem of determining whether or not $M(\theta)$ is sufficient for all $\theta \in \Theta$ is not trivial, in general. Although we acknowledge the difficulty of this problem, solving it is not the focus of this work. We note that, in practice, satisfaction of the stronger condition that $M(\theta)$ be positive semi-definite for all $\theta \in \Theta$ is easier to verify. Even under this condition, though, the procedures we present still allow one to solve several important parametric problems for which there was previously no available solution technique. We discuss a variety of such problems in the next several paragraphs, and we note that for each of these problems, $M(\theta)$ will always be positive semi-definite for all $\theta \in \Theta$.

Since mpLCP as in (1.1) has not yet been studied for $k > 1$, we briefly discuss some of the problems that can be solved using the methods proposed in this work. It is well known that both linear programs (LPs) and convex quadratic programs (QPs) can be reformulated as LCPs. Thus, (1.1) encompasses two very important classes of problems:

(i) Multiparametric LP (mpLP):

$$\begin{aligned} \min_x \quad & c(\theta)^\top x \\ \text{s.t.} \quad & A(\theta)x \le b(\theta) \end{aligned} \tag{1.4}$$

(ii) Multiparametric (convex) QP (mpQP):

$$\begin{aligned} \min_x \quad & \tfrac{1}{2}x^\top Q(\theta)x + c(\theta)^\top x \\ \text{s.t.} \quad & A(\theta)x \le b(\theta) \end{aligned} \tag{1.5}$$

Parametric LP with $A(\theta) = A$ and parametric QP with $A(\theta) = A$ and $Q(\theta) = Q$ have been studied extensively. Pistikopoulos et al. [43] provided an excellent survey of the literature for the case in which $k > 1$. Parametric LP with $A(\theta) \neq A$ has also been studied for quite some time. Perhaps the earliest work is due to Courtillot [15]. Solution techniques for various special cases of parametric LP with $k = 1$ and $A(\theta) \neq A$ are presented in works such as Barnett [4], Dent et al. [16], Finkelstein and Gumenok [19], Kim [34], Willner [56], and Filar et al. [18]. The works of Väliaho [49] and Khalilpour and Karimi [33] introduced methods for solving (1.4) with $k = 1$ and $A(\theta) \neq A$. Kolev and Skalna [35] studied (1.4) in the context of Interval Linear Programming (see [30], for example). Charitopoulos et al. [10] studied (1.4) in general. We point out that while both Kolev and Skalna [35] and Charitopoulos et al. [10] provided solution strategies for (1.4), each requires that

Θ be a bounded hypercube. The method we present in this work is also capable of solving (1.4), but for the more general case in which Θ is a bounded semi-algebraic set.

In addition to the works cited in Pistikopoulos et al. [43], parametric QP with $A(\theta) = A$ and $Q(\theta) = Q$ is considered in Ghaffari-Hadigheh et al. [23] and Ghaffari-Hadigheh et al. [24], particularly for the case in which $k = 2$. Solution procedures for various versions of (1.5) in which $A(\theta) = A$ are discussed in Bank et al. [3]. Additionally, solution techniques are presented for (1.5) with $k = 1$, $A(\theta) \neq A$, and $Q(\theta) \neq Q$ in Ritter [44], Väliaho [50], and Jonker et al. [32]. We are unaware of any work that provides a method for solving (1.5) with $k > 1$, $A(\theta) \neq A$, and $Q(\theta) \neq Q$. The method we present is capable of solving this problem.

Another class of problems that can be reformulated and solved using (1.1) is multiobjective programming problems having linear and/or convex quadratic objective functions and linear constraints. This is due to the common method of solving multiobjective problems using scalarization techniques that transform the problem into a single objective problem by introducing one or more parameters. For a detailed discussion on multiobjective programming and the various scalarization techniques available, see Ehrgott [17]. Multiobjective programs in which all objective functions are linear have been widely studied and can be solved efficiently using the multiobjective simplex method (Ehrgott [17]). Efficient methods have also been proposed for problems with one convex quadratic objective and one or more linear objectives, see, for example, Hirschberger et al. [29], Steuer et al. [47], and Hirschberger et al. [28]. Goh and Yang [25] presented a method for solving multiobjective problems with two or more convex quadratic objectives, though they impose a few minor restrictions. The work we present here serves as an alternative method for solving multiobjective programs with any number of linear and/or convex quadratic objectives, without restriction. Recently, Jayasekara et al. [31] exploited the methodology presented herein (as found in the preliminary version [1]) in order to solve multiobjective portfolio optimization problems. Utilization of the novel ideas included here permitted the authors to solve problems containing multiple quadratic risk functions—a class of problems that had previously been unsolved.

The procedure for solving mpLCP (1.1) proposed in this work consists of two phases. However, we discuss phase 2 prior to phase 1 because the problem solved in the first phase is a special case of the problem solved during the second phase. As a result, we proceed with the rest of this work in the following fashion. Chapter 2 contains a review of relevant information on LCP problems, including some interesting geometrical properties. In Chapter 3 we discuss algebraic properties of invariancy regions. The theory and methodology for our method of solving mpLCP are presented in Chapters 4 and 5. The former focuses on phase 2, while the latter focuses on phase 1. In Chapter 6 we point out some of the difficulties that can arise within certain instances of mpLCP and explain the strategies we employ in order to appropriately deal with these difficulties. Specifically, in Section 6.1 we introduce two examples that highlight the importance of Assumptions 1.1 and 1.2, and in Section 6.2 we discuss the uniqueness of partitions of Θ. The results of a computational experiment and a discussion on the performance of the proposed algorithms are presented in Chapter 7. Finally, in Chapter 8 we make concluding remarks, including a discussion on proposed future work. We also include two appendices, Appendices A and B, in which we provide tables that are important for the examples we discuss in Chapters 4, 5, and 6.

Before continuing to Chapter 2, we pause briefly to highlight some of the major differences between the work presented here and that of Adelgren and Wiecek [2]. The major difference, of course, stems from the fact that parameters are now permitted in the M matrix, whereas they were not in the previous work. This seemingly slight change has far-reaching implications, though. For example, in the previous work all invariancy regions were guaranteed to be polyhedral, and all work necessary for their discovery could be done through the use of linear programming. When the M matrix is permitted to be parametric, however, invariancy regions are defined by a set of polynomial inequalities and are not guaranteed to be convex, or even connected. Moreover, discovery of such regions relies heavily on the solution of polynomial optimization problems. Note that while this difference in the structure of the invariancy regions between the two classes of problems will be quite noticeable to anyone wishing to apply the methodology presented herein, it stems from a major difference in the underlying problem structure that will not necessarily be apparent to the end user. As will be explained in more detail later, invariancy regions are the result of applying a special mapping from $\mathbb{R}^h$ into $\mathbb{R}^k$ to the intersection of a cone and a hyperplane (or a subset of a hyperplane, specified by Θ). Specifically, $q(\theta)$ identifies a k-dimensional hyperplane in $\mathbb{R}^h$, where each $\theta \in \Theta$ is associated with a unique point in the hyperplane, and the columns of the M matrix, together with the h standard basis column vectors, identify a set of cones in $\mathbb{R}^h$. Each of these cones that has nonempty intersection with the hyperplane specified by $q(\theta)$ then identifies a nonempty invariancy region. Specifically, the invariancy region associated with a given h-dimensional cone contains each $\theta \in \Theta$ for which $q(\theta)$ is within the given cone. Clearly, in the case in which M is not permitted to be parametric, the orientation of each cone is fixed, and thus determining the set of $\theta \in \Theta$ for which $q(\theta)$ is in the cone is relatively straightforward. However, when M is permitted to be parametric, the orientation of the cones identified by M may be different for each $\theta \in \Theta$, making it is far more challenging to develop the theory necessary to determine the set of $\theta \in \Theta$ for which $q(\theta)$ is in a given cone. More details on the underlying problem structure of mpLCP are given in Chapter 2.

While there are many differences between the work we present here and that of Adelgren and Wiecek [2], there are also some important similarities. Perhaps the most important of these is the similarity in methodology presented in both works. In order to obtain a partition of Θ, in both cases a two-phase procedure is proposed in which phase 1 is used to obtain an initial invariancy region of dimension k, and phase 2 is used to iteratively generate additional k-dimensional invariancy regions until a partition of Θ has been obtained. In both cases, we are able to show that this can be done through consideration of only k and $(k-1)$-dimensional invariancy regions. Details are presented in Chapters 4 and 5.

Chapter 2
Background on mpLCP

This chapter is divided into three sections. In Section 2.1 we present important definitions and some notation. In Section 2.2 we provide a detailed discussion on invariancy regions and the properties of these sets. Finally, in Section 2.3 we study the geometry of mpLCP. As many of the concepts we make use of throughout this work are new and rather advanced, we introduce here two examples of small instances of mpLCP (1.1) that we will refer back to at various locations throughout this work.

Example 2.1

$$
\begin{aligned}
w - \begin{bmatrix} 0 & 0 & 1 & 3 & -5 \\ 0 & 0 & 2 & 2 & 2 \\ -1 & -2 & 2\theta_1 - \theta_2 + 4 & \theta_1 - 2\theta_2 + 3 & 3\theta_1 + 4\theta_2 - 2 \\ -3 & -2 & \theta_1 - 2\theta_2 + 3 & -\theta_1 + \theta_2 + 4 & 3\theta_1 + 4\theta_2 - 3 \\ 5 & -2 & 3\theta_1 + 4\theta_2 - 2 & 3\theta_1 + 4\theta_2 - 3 & -\theta_2 + 3 \end{bmatrix} z &= \begin{bmatrix} 3 \\ -\theta_1 - 2 \\ 0 \\ 0 \\ 0 \end{bmatrix} \\
w^\top z &= 0 \\
w, z &\geq 0
\end{aligned}
\tag{2.1}
$$

Note here that $h = 5$ and $k = 2$. Assume that $\Theta = \{\theta \in \mathbb{R}^2 : \theta \geq 0, \theta_1 + \theta_2 \leq 1\}$. It is easy to verify for this example that $M(\theta)$ is sufficient for all $\theta \in \Theta$ since $M(\theta)$ is actually positive semi-definite for each $\theta \in \Theta$.

N. Adelgren, *Advancing Parametric Optimization*, SpringerBriefs in Optimization,
https://doi.org/10.1007/978-3-030-61821-6_2

Example 2.2

$$w - \begin{bmatrix} 0 & 0 & -2 & -1 \\ 0 & 0 & -5 & \theta_1 + 7 \\ 1 & 3 & 0 & 0 \\ 1 & -\theta_1 - 5 & 0 & 0 \end{bmatrix} z = \begin{bmatrix} -\theta_2 - 1 \\ \theta_1 - \theta_2 - 1 \\ -18\theta_2 - 34 \\ -9\theta_2 - 17 \end{bmatrix}$$
$$w^\top z = 0$$
$$w, z \geq 0 \tag{2.2}$$

For this example, $h = 4$ and $k = 2$. Assume that $\Theta = [-3, 1]^2$. It can be verified that $M(\theta)$ is sufficient for all $\theta \in \Theta$ by using Lemma 1.1.

2.1 Preliminaries

In this section we provide definitions and notation that are needed throughout this work. Note that several of the definitions are slightly modified from well-known definitions for LCP. Given an instance of mpLCP as in (1.1), define the matrix $G(\theta) := \begin{bmatrix} I & -M(\theta) \end{bmatrix}$ and the vector $v := \begin{bmatrix} w \\ z \end{bmatrix}$, where $G(\theta) \in \mathbb{R}^{h \times h} \times \Theta^{h \times h}$ and $v \in \mathbb{R}^{2h}$. Then $G(\theta)_{i\bullet}$ and $G(\theta)_{\bullet j}$ represent the ith row and jth column of $G(\theta)$, respectively. Similarly, given appropriate index sets $\mathcal{I}$ and $\mathcal{J}$, we use $G(\theta)_{\mathcal{I}\bullet}$, $G(\theta)_{\bullet\mathcal{J}}$, and $G(\theta)_{\mathcal{I}\mathcal{J}}$ to denote the matrix formed by the rows of $G(\theta)$ indexed by $\mathcal{I}$, the matrix formed by the columns of $G(\theta)$ indexed by $\mathcal{J}$, and the matrix formed by the elements of the rows indexed by $\mathcal{I}$ that are in the columns indexed by $\mathcal{J}$, respectively. Now, define $\mathcal{E} := \{1, \ldots, 2h\}$.

Definition 2.1 Given $\mathcal{I} \subset \mathcal{E}$, $\theta \in \Theta$ is said to be *rank preserving over* $\mathcal{I}$ if $rank\,(G(\theta)_{\bullet\mathcal{I}}) = |\mathcal{I}|$.

We then let

$$\Theta_{\mathcal{I}} := \{\theta \in \Theta : rank\,(G(\theta)_{\bullet\mathcal{I}}) = |\mathcal{I}|\} \tag{2.3}$$

be the set of all rank preserving $\theta \in \Theta$ for $\mathcal{I} \subset \mathcal{E}$.

Definition 2.2 A set $\mathcal{B} \subset \mathcal{E}$ is a *basis* if $|\mathcal{B}| = h$ and $\Theta_{\mathcal{B}} \neq \emptyset$. $\mathcal{N} := \mathcal{E} \setminus \mathcal{B}$ is the *complement* of $\mathcal{B}$.

Definition 2.3 Together, a basis $\mathcal{B}$ and its complement $\mathcal{N}$ specify the sets of *basic* and *nonbasic* variables: $v_{\mathcal{B}} := \{v_i : i \in \mathcal{B}\}$ and $v_{\mathcal{N}} := \{v_i : i \in \mathcal{N}\}$, respectively.

Definition 2.4 Given a basis $\mathcal{B}$, for each $\theta \in \Theta_{\mathcal{B}}$, $v_{\mathcal{B}}(\theta) = G(\theta)^{-1}_{\bullet\mathcal{B}} q(\theta)$, $v_{\mathcal{N}}(\theta) = 0$ is a solution to the linear system

$$G(\theta)v = q(\theta).$$

For each $\theta \in \Theta_{\mathcal{B}}$, the solution $\left(\nu_{\mathcal{B}}(\theta), \nu_{\mathcal{N}}(\theta)\right)$ is called a *basic solution.*

Definition 2.5 A basis $\mathcal{B}$ is *complementary* if $\left|\{i, i+h\} \cap \mathcal{B}\right| = 1$ for each $i \in \{1, \ldots, h\}$.

We are now prepared to provide the definition of an *invariancy region*. Although we do provide the definition here, we note that Section 2.2 is devoted to a more detailed discussion on these sets.

Let a complementary basis $\mathcal{B}$ be given for which there exists $\theta \in \Theta_{\mathcal{B}} \subseteq \Theta$ such that (i) $\nu_{\mathcal{B}}(\theta) = G(\theta)^{-1}_{\bullet\mathcal{B}} q(\theta) \geq 0$ and (ii) $\nu_{\mathcal{N}}(\theta) = 0$. Then since $\nu = \begin{bmatrix} w \\ z \end{bmatrix}$, for all $\theta \in \Theta_{\mathcal{B}}$ satisfying (i) and (ii) above, the basic solution $\left(\nu_{\mathcal{B}}(\theta), \nu_{\mathcal{N}}(\theta)\right)$ satisfies (1.1) and defines the solution vectors $w(\theta)$ and $z(\theta)$ for mpLCP. Now, because there may exist one set of solution vectors of this form for each complementary basis, there may exist one invariancy region for each complementary basis.

Definition 2.6 Given a complementary basis $\mathcal{B}$, its associated *invariancy region* $\mathcal{IR}_{\mathcal{B}}$ is the set

$$\mathcal{IR}_{\mathcal{B}} := \left\{\theta \in \Theta_{\mathcal{B}} : G(\theta)^{-1}_{\bullet\mathcal{B}} q(\theta) \geq 0\right\}. \tag{2.4}$$

Definition 2.7 A complementary basis $\mathcal{B}$ is *feasible* to (1.1) if $\mathcal{IR}_{\mathcal{B}} \neq \emptyset$.

Each invariancy region is a possibly non-convex subset of Θ. For every feasible complementary basis (from here abbreviated f.c.b.) $\mathcal{B}$, the function $\nu_{\mathcal{B}}(\theta) = G(\theta)^{-1}_{\bullet\mathcal{B}} q(\theta)$, $\nu_{\mathcal{N}}(\theta) = 0$ is a solution to (1.1) for all $\theta \in \mathcal{IR}_{\mathcal{B}}$. Furthermore, as will be discussed in more detail in Section 2.3, when $M(\theta)$ is sufficient for each $\theta \in \Theta$, there is an onto correspondence between solutions to mpLCP and f.c.b.'s, i.e., for each $\theta \in \Theta$, there exists a f.c.b. $\mathcal{B}$ for which $(\nu_{\mathcal{B}}, \nu_{\mathcal{N}})$ solves mpLCP at θ. As a result, the primary focus of this work is to present a procedure for determining a piecewise solution to (1.1) by partitioning Θ into a set of invariancy regions, where each invariancy region is associated with a unique f.c.b. Note that we do not claim that this partition is unique, only that given a partition, each invariancy region is associated with a unique f.c.b.

2.2 Invariancy Regions

In order to gain a deeper understanding of invariancy regions, we return to Examples 2.1 and 2.2. For Example 2.1, it is not difficult to verify that for $\theta = \begin{bmatrix} \frac{1}{5} \\ \frac{2}{5} \end{bmatrix}$, the basis $\mathcal{B}_0^{2.1} = \{w_1, z_2, z_3, z_4, z_5\}$ is feasible (see Table A.7 in Appendix A). Similarly, for Example 2.2, the basis $\mathcal{B}_0^{2.2} = \{w_1, w_2, w_3, w_4\}$ is feasible for

$\theta = \begin{bmatrix} -2 \\ -2 \end{bmatrix}$ (see Table B.1 in Appendix B). Note that for ease of understanding, we are representing $\mathcal{B}_0^{2.1}$ and $\mathcal{B}_0^{2.2}$ as the sets of variables given by the indices in each basis, rather than the sets of indices themselves. We will often represent bases this way throughout this work. Observe the invariancy regions for $\mathcal{B}_0^{2.1}$ and $\mathcal{B}_0^{2.2}$:

$$\mathcal{IR}_{\mathcal{B}_0^{2.1}} = \left\{ \theta \in \Theta_{\mathcal{B}_0^{2.1}} : \begin{array}{c} \frac{3\theta_1^3+18\theta_1^2\theta_2-49\theta_1^2-75\theta_1\theta_2^2+148\theta_1\theta_2+68\theta_1+96\theta_2^2-16\theta_2-76}{2(-3\theta_1^2+8\theta_1\theta_2+19\theta_1+41\theta_2^2-24\theta_2-22)} \geq 0 \\ \frac{-(\theta_1+2)(9\theta_1^3-9\theta_1^2\theta_2-33\theta_1^2-87\theta_1\theta_2^2+21\theta_1\theta_2+22\theta_1-59\theta_2^3+13\theta_2^2+50\theta_2+5)}{4(-3\theta_1^2+8\theta_1\theta_2+19\theta_1+41\theta_2^2-24\theta_2-22)} \geq 0 \\ \frac{(\theta_1+2)(-6\theta_1^2-\theta_1\theta_2+11\theta_1+15\theta_2^2-16\theta_2+1)}{2(-3\theta_1^2+8\theta_1\theta_2+19\theta_1+41\theta_2^2-24\theta_2-22)} \geq 0 \\ \frac{(\theta_1+2)(3\theta_1^2+8\theta_1\theta_2-\theta_1+5\theta_2^2+5\theta_2-11)}{2(-3\theta_1^2+8\theta_1\theta_2+19\theta_1+41\theta_2^2-24\theta_2-22)} \geq 0 \\ \frac{(\theta_1+2)(9\theta_1-13\theta_2+\theta_1\theta_2+21\theta_2^2-12)}{2(-3\theta_1^2+8\theta_1\theta_2+19\theta_1+41\theta_2^2-24\theta_2-22)} \geq 0 \end{array} \right\}$$

$$\mathcal{IR}_{\mathcal{B}_0^{2.2}} = \left\{ \theta \in \Theta_{\mathcal{B}_0^{2.2}} : \begin{array}{c} -\theta_2 - 1 \geq 0 \\ \theta_1 - \theta_2 - 1 \geq 0 \\ -18\theta_2 - 34 \geq 0 \\ -9\theta_2 - 17 \geq 0 \end{array} \right\}$$

The following propositions arise from conclusions one may draw from observing an invariancy region such as those above.

Proposition 2.1 *Given a f.c.b. $\mathcal{B}$, the invariancy region $\mathcal{IR}_{\mathcal{B}}$ is defined by a set of rational inequalities, all having the same denominator.*

Proof Recognize that $G(\theta)_{\bullet\mathcal{B}}^{-1} = \dfrac{Adj\,(G(\theta)_{\bullet\mathcal{B}})}{det\,(G(\theta)_{\bullet\mathcal{B}})}$, where $Adj(\cdot)$ and $det(\cdot)$ represent the matrix adjoint and determinant, respectively. Thus, the result follows from (2.4). □

Proposition 2.1 shows that invariancy regions have a relatively nice structure. However, this structure is actually better than it seems. This is shown in the following lemma and the subsequent proposition.

Lemma 2.1 *Given a f.c.b. $\mathcal{B}$, one of the following must hold:*

1. $det\,(G(\theta)_{\bullet\mathcal{B}}) \geq 0$ *for all* $\theta \in \Theta$
2. $det\,(G(\theta)_{\bullet\mathcal{B}}) \leq 0$ *for all* $\theta \in \Theta$.

Proof Recall that we assume that $M(\theta)$ is sufficient for each $\theta \in \Theta$. It is well known that sufficient matrices are also $\mathbf{P_0}$ matrices, i.e., their principal minors are all nonnegative (see, for example, Cottle et al. [12] or Väliaho [52]). Hence, recognize that for any $n \in \{1, \ldots, h\}$, the nth-order principal minors of $-M(\theta)$ are each (i) nonnegative for all $\theta \in \Theta$ if n is even or (ii) nonpositive for all $\theta \in \Theta$ if n is odd. Notice that there exists $J \subseteq \{0, \ldots, k\}$ such that $G(\theta)_{\bullet\mathcal{B}}$ is obtained from

$-M(\theta)$ by replacing column $(-M(\theta))_{\bullet j}$ with $I_{\bullet j}$ for each $j \in J$. Thus, if $|J| = h$, $det\,(G(\theta)_{\bullet\mathcal{B}}) = det(I) = 1$ and if $|J| \neq h$, $det\,(G(\theta)_{\bullet\mathcal{B}})$ is equal to a $(h - |J|)$th-order principal minor of $-M(\theta)$. Therefore, if $(h - |J|)$ is even, condition (1) holds, and if $(h - |J|)$ is odd, condition (2) holds. □

Note that the case in which $det\,(G(\theta)_{\bullet\mathcal{B}}) = 0$ for some $\theta \in \Theta$ is not concerning because we observe from (2.3) and (2.4) that in this case $\theta \notin \mathcal{IR}_{\mathcal{B}}$.

Proposition 2.2 *For every f.c.b. $\mathcal{B}$, the invariancy region $\mathcal{IR}_{\mathcal{B}}$ is defined by a set of polynomial inequalities.*

Proof From Proposition 2.1, we know that all defining inequalities of $\mathcal{IR}_{\mathcal{B}}$ are given by rational functions whose denominators equal $det\,(G(\theta)_{\bullet\mathcal{B}})$. By Lemma 2.1, we know that either (1) $det\,(G(\theta)_{\bullet\mathcal{B}}) \geq 0$ for all $\theta \in \Theta$ or (2) $det\,(G(\theta)_{\bullet\mathcal{B}}) \leq 0$ for all $\theta \in \Theta$. Therefore, since for all $\theta \in \Theta_{\mathcal{B}}$, we have $det\,(G(\theta)_{\bullet\mathcal{B}}) \neq 0$, under condition (1), an equivalent formulation for any invariancy region can be given by ensuring that the numerator of each defining rational inequality is nonnegative. Similarly, under condition (2), an equivalent formulation for any invariancy region can be given by ensuring that the numerator of each defining rational inequality is nonpositive. □

As a consequence of Lemma 2.1, for each f.c.b. $\mathcal{B}$, we define the following:

$$g_{\mathcal{B}} := \begin{cases} 1 & \text{if } det\,(G(\theta)_{\bullet\mathcal{B}}) \geq 0 \text{ for all } \theta \in \Theta \\ -1 & \text{if } det\,(G(\theta)_{\bullet\mathcal{B}}) \leq 0 \text{ for all } \theta \in \Theta \end{cases} \tag{2.5}$$

Then for each f.c.b. $\mathcal{B}$, the associated invariancy region can be expressed with polynomial defining inequalities as

$$\mathcal{IR}_{\mathcal{B}} := \{\theta \in \Theta_{\mathcal{B}} : g_{\mathcal{B}} Adj\,(G(\theta)_{\bullet\mathcal{B}})\, q(\theta) \geq 0\}. \tag{2.6}$$

We now recall the following definition from [8].

Definition 2.8 A *semi-algebraic subset* of $\mathbb{R}^n$ is a subset of the form

$$\bigcup_{i=1}^{s} \bigcap_{j=1}^{r_i} \left\{x \in \mathbb{R}^n : f_{i,j}(x) \star_{i,j} 0\right\}, \tag{2.7}$$

where, for each $i \in \{1, \ldots, s\}$ and $j \in \{1, \ldots, r_i\}$, $f_{i,j}$ is a polynomial function on $\mathbb{R}^n$ and $\star_{i,j}$ represents either "$=$" or "$<$."

We note that a subset of $\mathbb{R}^n$ in the form of (2.7) is simply called *algebraic* whenever $\star_{i,j}$ represents "$=$" for every i, j pair. Now, from Proposition 2.2, we are able to make the following observation.

Observation 2.1 *Every invariancy region is a semi-algebraic subset of Θ.*

Return to Examples 2.1 and 2.2 and observe the invariancy regions for $\mathcal{B}_0^{2.1}$ and $\mathcal{B}_0^{2.2}$ expressed as semi-algebraic sets:

$$\mathcal{IR}_{\mathcal{B}_0^{2.1}} = \left\{ \theta \in \Theta_{\mathcal{B}_0^{2.1}} : \begin{array}{c} 3\theta_1^3 + 18\theta_1^2\theta_2 - 49\theta_1^2 - 75\theta_1\theta_2^2 + 148\theta_1\theta_2 + 68\theta_1 + 96\theta_2^2 - 16\theta_2 - 76 \le 0 \\ -(\theta_1 + 2)(9\theta_1^3 - 9\theta_1^2\theta_2 - 33\theta_1^2 - 87\theta_1\theta_2^2 + 21\theta_1\theta_2 + 22\theta_1 - 59\theta_2^3 + 13\theta_2^2 + 50\theta_2 + 5) \le 0 \\ (\theta_1 + 2)(-6\theta_1^2 - \theta_1\theta_2 + 11\theta_1 + 15\theta_2^2 - 16\theta_2 + 1) \le 0 \\ (\theta_1 + 2)(3\theta_1^2 + 8\theta_1\theta_2 - \theta_1 + 5\theta_2^2 + 5\theta_2 - 11) \le 0 \\ (\theta_1 + 2)(9\theta_1 - 13\theta_2 + \theta_1\theta_2 + 21\theta_2^2 - 12) \le 0 \end{array} \right\} \tag{2.8}$$

$$\mathcal{IR}_{\mathcal{B}_0^{2.2}} = \left\{ \theta \in \Theta_{\mathcal{B}_0^{2.2}} : \begin{array}{c} -\theta_2 - 1 \ge 0 \\ \theta_1 - \theta_2 - 1 \ge 0 \\ -18\theta_2 - 34 \ge 0 \\ -9\theta_2 - 17 \ge 0 \end{array} \right\} \tag{2.9}$$

2.3 Geometry of the mpLCP

In this section we discuss some of the properties of the (mp)LCP problem that we will need in order to establish an algorithm for partitioning Θ. We now give several more definitions and properties needed for this discussion.

Definition 2.9 For an index $i \in \mathcal{E}$, the *complementary index* of i is $\bar{\imath} := \begin{cases} i + h & \text{if } i \le h \\ i - h & \text{otherwise.} \end{cases}$

Similarly, given $\mathcal{I} \subseteq \mathcal{E}$, the set of all complementary indices of elements in $\mathcal{I}$ is denoted as $\bar{\mathcal{I}}$.

Definition 2.10 A set $\mathcal{J} \subset \mathcal{E}$ is called *complementary* if $i \in \mathcal{J} \Rightarrow \bar{\imath} \notin \mathcal{J}$.

Given an arbitrary matrix Q, we use the notation $cone(Q)$ to represent the set of all nonnegative combinations of the columns of Q, i.e., $cone(Q) = \{\tau : \tau = Qx, x \ge 0\}$. If Q is nonsingular, $cone(Q) = \{\tau : Q^{-1}\tau \ge 0\}$ is an equivalent formulation.

Definition 2.11 (Modified from [26]) Let $\{a_i\}_{i=1}^k$ be a given set of vectors in $\mathbb{R}^n$. A cone $C \subseteq \mathbb{R}^n$ is called a *convex polyhedral cone* if it has the form

$$C = \{x : a_i^\top x \le 0, i = 1, ..., k\}.$$

In other words, a polyhedral cone is the intersection of finitely many half-spaces passing through the origin.

Definition 2.12 For any complementary set $\mathcal{J}$ and any $\theta \in \Theta$, the set $\mathcal{C}_{\mathcal{J}}(\theta) := cone\left(G(\theta)_{\bullet\mathcal{J}}\right)$ is called a *parametric complementary cone* with respect to the matrix $M(\theta)$.

Observation 2.2 *For any complementary set $\mathcal{J}$ and any $\theta \in \Theta$, the parametric complementary cone $\mathcal{C}_{\mathcal{J}}(\theta)$ is a convex polyhedral cone.*

Definition 2.13 A parametric complementary cone $\mathcal{C}_{\mathcal{J}}(\theta)$ is called *full dimensional* if $dim(\mathcal{C}_{\mathcal{J}}(\theta)) = h$, i.e., if $rank\left(G(\theta)_{\bullet\mathcal{J}}\right) = h$.

Proposition 2.3 *A parametric complementary cone $\mathcal{C}_{\mathcal{J}}(\theta)$ is full dimensional if and only if $\mathcal{J}$ is a complementary basis and $\theta \in \Theta_{\mathcal{J}}$.*

Proof ($\Rightarrow$): Since $\mathcal{C}_{\mathcal{J}}(\theta)$ is a parametric complementary cone, $\mathcal{J}$ is complementary, and so $|\mathcal{J}| = h$. Since $\mathcal{C}_{\mathcal{J}}(\theta)$ is full dimensional, $rank\left(G(\theta)_{\bullet\mathcal{J}}\right) = h$ and so $\theta \in \Theta_{\mathcal{J}}$.
($\Leftarrow$): Since $\mathcal{J}$ is a complementary basis and $\theta \in \Theta_{\mathcal{J}}$, we have $rank\left(G(\theta)_{\bullet\mathcal{J}}\right) = h$. Thus, $dim\left(\mathcal{C}_{\mathcal{J}}(\theta)\right) = h$. □

Observation 2.3 *For any complementary basis $\mathcal{B}$ and $\theta \in \Theta_{\mathcal{B}}$, the parametric complementary cone $\mathcal{C}_{\mathcal{B}}(\theta)$ is a full dimensional convex polyhedral cone. Hence, for any $\mathcal{J} \subseteq \mathcal{B}$, the set cone $\left(G(\theta)_{\bullet(\mathcal{B}\setminus\mathcal{J})}\right)$ is a face of $\mathcal{C}_{\mathcal{B}}(\theta)$. Furthermore, for any $i \in \mathcal{B}$, the set cone $\left(G(\theta)_{\bullet(\mathcal{B}\setminus\{i\})}\right)$ is a facet of $\mathcal{C}_{\mathcal{B}}(\theta)$.*

Definition 2.14 For distinct complementary bases $\mathcal{B}_1$ and $\mathcal{B}_2$ and fixed $\theta \in \Theta$, the parametric complementary cones $\mathcal{C}_{\mathcal{B}_1}(\theta)$ and $\mathcal{C}_{\mathcal{B}_2}(\theta)$ are called *adjacent* if $dim\left(\mathcal{C}_{\mathcal{B}_1}(\theta) \cap \mathcal{C}_{\mathcal{B}_2}(\theta)\right) = h - 1$. In this case, the bases $\mathcal{B}_1$ and $\mathcal{B}_2$ are also called adjacent.

For a complementary basis $\mathcal{B}$ and $\theta \in \Theta$, the associated parametric complementary cone is

$$\mathcal{C}_{\mathcal{B}}(\theta) = \left\{\tau \in \mathbb{R}^h : G(\theta)^{-1}_{\bullet\mathcal{B}}\tau \geq 0\right\} \tag{2.10}$$

Thus, given some $\theta \in \Theta$, $\mathcal{C}_{\mathcal{B}}(\theta)$ defines the set of all vectors for which the basis $\mathcal{B}$ is feasible. Similarly, $\mathcal{K}(M(\theta)) := \cup_{\text{feasible bases } \mathcal{B}} \mathcal{C}_{\mathcal{B}}(\theta)$ defines the set of all vectors for which any basis is feasible.

Note the close relationship between $\mathcal{IR}_{\mathcal{B}}$ and $\mathcal{C}_{\mathcal{B}}(\theta)$. In order to understand this relationship further, we now consider more closely the vector $q(\theta)$. Since every element of the vector $q(\theta)$ is an affine function of θ, there exists a vector $q \in \mathbb{R}^h$ and a matrix $\triangle Q \in \mathbb{R}^{h\times k}$ such that $q(\theta) = q + \triangle Q\theta$. Throughout this work, we assume that the matrix $\triangle Q$ is of full column rank. In the case when $\triangle Q$ is not of full column rank, there exists a "reduced" mpLCP as described in Section 2 of [2] that can be solved in place of the original mpLCP. Consider the following definition.

Definition 2.15 For a given f.c.b. $\mathcal{B}$, the *invariant domain of* $\mathcal{B}$ is the union over $\theta \in \Theta$ of intersections of the vectors $q(\theta)$ with the parametric complementary cones $\mathcal{C}_{\mathcal{B}}(\theta)$ and is denoted as $\mathcal{ID}_{\mathcal{B}}$. Hence,

$$\mathcal{ID}_{\mathcal{B}} := \bigcup_{\theta \in \Theta} (\mathcal{C}_{\mathcal{B}}(\theta) \cap q(\theta)) = \left\{\tau : G(\theta)^{-1}_{\bullet \mathcal{B}} \tau \geq 0, \tau = q(\theta), \theta \in \Theta\right\}. \quad (2.11)$$

Under the assumption that $\triangle Q$ has full column rank, $q(\theta)$ can be viewed as a bijective function with inverse $q^{-1}(\cdot)$. The following two observations result.

Observation 2.4 *For any f.c.b.* $\mathcal{B}$*, we have* $\mathcal{ID}_{\mathcal{B}} = q(\mathcal{IR}_{\mathcal{B}})$ *and* $\mathcal{IR}_{\mathcal{B}} = q^{-1}(\mathcal{ID}_{\mathcal{B}})$.

Observation 2.5 *For each* $i \in \mathcal{E}$*, the inequality* $\left(G(\theta)^{-1}_{\bullet \mathcal{B}}\right)_{i \bullet} \tau \geq 0$ *is redundant in* $\mathcal{ID}_{\mathcal{B}}$ *if and only if* $\left(G(\theta)^{-1}_{\bullet \mathcal{B}}\right)_{i \bullet} q(\theta) \geq 0$ *is redundant in* $\mathcal{IR}_{\mathcal{B}}$*. (Note that a defining inequality of a set is referred to as redundant if removing the inequality from the set's definition does not change the set.)*

Notice that, given a f.c.b. $\mathcal{B}$, these observations make clear the relationship existing between the invariancy region $\mathcal{IR}_{\mathcal{B}}$ and the parametric complementary cone $\mathcal{C}_{\mathcal{B}}(\theta)$. Observe the following definition that we employ when dealing with convex sets and refer to in the paragraph below.

Definition 2.16 For an arbitrary convex set S, the relative interior of S is the set $relint(S) := \{s \in S : \exists \epsilon > 0,\ B_\epsilon(s) \cap aff(S) \subseteq S\}$, where $B_\epsilon(s)$ is the ball of radius ϵ centered at s and $aff(S)$ is the affine hull of S. Said differently, $relint(S)$ is the intersection of all affine sets containing S.

Briefly consider non-parametric LCP. In this context, we use the notations M and $\mathcal{K}(M)$ to denote the non-parametric counterparts of $M(\theta)$ and $\mathcal{K}(M(\theta))$ and the term *complementary cone* to refer to the non-parametric counterpart of a parametric complementary cone. Given an instance of LCP, the properties of the matrix M strongly impact the structure of the complementary cones. Clearly this extends to mpLCP and shows that the properties of $M(\theta)$ influence the structure of the parametric complementary cones and invariancy regions. In this work, the classes of column sufficient matrices and Q_0 matrices are particularly useful. Recall that we defined column sufficiency in Definition 1.1. For Q_0 matrices, we provide the following definition, as it appears in [42].

Definition 2.17 The square matrix M is said to be a Q_0 matrix if $\mathcal{K}(M)$ is a convex cone.

We now recall the following important property of column sufficient matrices, shown in [14]: if M is column sufficient, all complementary cones have disjoint relative interiors, i.e., the complementary cones partition $\mathcal{K}(M)$. The theory we develop throughout the rest of this work and the algorithms we present in Chapters 4 and 5 rely heavily on these properties of column sufficient and Q_0 matrices. As a result, the algorithms we propose are designed for instances of mpLCP in which $M(\theta)$ is both Q_0 and column sufficient for all $\theta \in \Theta$. The class of sufficient matrices is the largest class of matrices that is known to be a subset of both the class of column sufficient matrices and the class of Q_0 matrices.

We now proceed to Chapter 3 where we delve deeper into the details of invariancy regions and develop the theory that shows that Θ can, in fact, be partitioned whenever Assumptions 1.1 and 1.2 are satisfied. Moreover, the theory developed in Chapter 3 allows us to establish properties of invariancy regions that we then exploit in order to specify the structure of a partition of Θ.

Chapter 3
Algebraic Properties of Invariancy Regions

Michael Burr of the School of Mathematical and Statistical Sciences at Clemson University has formulated the main conceptual ideas presented in Sections 3.1 *and* 3.2.

In this chapter we merge several concepts from the fields of Semi-algebraic Geometry, Algebraic Topology and Operations Research. We point out that although the details presented here are theoretically important and provide the necessary foundation for the development of the algorithms presented in Chapters 4 and 5, they are not necessary for the understanding of later chapters and may be skipped by the uninterested reader. Additionally, we note that the results presented in this chapter are not novel. In fact, for readers well versed in Semi-algebraic Geometry, many of these results are likely well known or even obvious if interpreted in the context of the cylindrical decomposition of semi-algebraic sets as discussed in Chapter 5 of [5]. Even so, we present these results in full detail but attempt to cater to readers with a background in Operations Research and/or mathematical optimization.

We note that this work is not the first to employ concepts from algebraic geometry in order to solve parametric systems or parametric optimization problems (see, for example, [20, 22, 37]). However, all works we are aware of that do employ algebraic geometry in order to solve parametric systems or parametric optimization problems do so directly through the use of cylindrical algebraic decomposition (CAD). While CAD is a powerful tool, it is also extremely complex and its use in practice is generally very computationally costly and requires the use of specialized computer algebra systems. Our use of algebraic geometry, on the other hand, is less direct and entirely theoretical. We show that invariancy regions share a special geometrical structure that can be exploited so that, given a starting k-dimensional invariancy region, each invariancy region present in a partition of Θ can be obtained from previously discovered invariancy regions by solving a series of nonlinear programs, without the use of CAD.

N. Adelgren, *Advancing Parametric Optimization*, SpringerBriefs in Optimization,
https://doi.org/10.1007/978-3-030-61821-6_3

3.1 Decomposition of the Parameter Space

We now continue our discussion of invariancy regions by decomposing Θ into subsets of interest. Consider an arbitrary element $M(\theta)_{ij}$ of the matrix $M(\theta)$. Recall that this element is an affine function of θ. It can therefore be represented as

$$M(\theta)_{ij} = \sum_{n=1}^{k} \alpha_{ijn}\theta_n + \beta_{ij}.$$

We now define U to be the support of $M(\theta)$ and V to be its complement. In other words,

$$U = \left\{n \in \{1, \ldots, k\} : \alpha_{ijn} \neq 0 \text{ for some } i, j \in \{1, \ldots, h\}\right\},$$

and $V = \{1, \ldots, k\} \setminus U$. Using these index sets, we denote $\phi := \operatorname{Proj}_{\mathbb{R}^U} \theta$ and $\upsilon := \operatorname{Proj}_{\mathbb{R}^V} \theta$. Thus, ϕ is the subvector of θ such that every element of ϕ is present in some element of $M(\theta)$ and υ is the subvector of θ such that no element of υ is present in any element of $M(\theta)$. We let $p = |U|$, and, consequently, $k - p = |V|$. Note that we allow for the case in which $p = k$. We do assume, however, that $p \neq 0$, since this is precisely the case dealt with in [2]. We now define

$$\begin{aligned} \Phi(\upsilon) &= \operatorname{Proj}_U(\operatorname{Proj}_V^{-1} \upsilon \cap \Theta) \\ &= \{\phi \in \mathbb{R}^p : \theta = (\phi, \upsilon) \in \Theta\} \end{aligned} \tag{3.1}$$

and

$$\begin{aligned} \Upsilon(\phi) &= \operatorname{Proj}_V(\operatorname{Proj}_U^{-1} \phi \cap \Theta) \\ &= \{\upsilon \in \mathbb{R}^{k-p} : \theta = (\phi, \upsilon) \in \Theta\}. \end{aligned} \tag{3.2}$$

$\Phi(\upsilon)$ can be interpreted as the set of attainable values of ϕ given a fixed υ, and similarly, $\Upsilon(\phi)$ can be interpreted as the set of attainable values of υ given a fixed ϕ. From these sets, we also define

$$\begin{aligned} \Phi &= \{\phi \in \mathbb{R}^p : \Upsilon(\phi) \neq \emptyset \text{ for some } \upsilon \in \mathbb{R}^{k-p}\} \\ &= \{\phi \in \mathbb{R}^p : \exists \upsilon \in \mathbb{R}^{k-p} \text{ s.t. } (\phi, \upsilon) \in \Theta\} \\ &= \operatorname{Proj}_U \Theta \end{aligned} \tag{3.3}$$

and

$$\begin{aligned} \Upsilon &= \{\upsilon \in \mathbb{R}^{k-p} : \Phi(\upsilon) \neq \emptyset \text{ for some } \phi \in \mathbb{R}^p\} \\ &= \{\upsilon \in \mathbb{R}^{k-p} : \exists \phi \in \mathbb{R}^p \text{ s.t. } (\phi, \upsilon) \in \Theta\} \\ &= \operatorname{Proj}_V \Theta, \end{aligned} \tag{3.4}$$

where Proj_S denotes the projection onto $S \subseteq \mathbb{R}^S$.

Now, given a basis $\mathcal{B}$ and some $\theta = (\phi, \upsilon) \in \Theta_{\mathcal{B}}$, we define the following subsets of the invariancy region $\mathcal{IR}_{\mathcal{B}}$:

$$\begin{aligned} \mathcal{IR}_{\mathcal{B}}(\upsilon) &= \{\phi \in \mathbb{R}^p : \theta = (\phi, \upsilon) \in \mathcal{IR}_{\mathcal{B}}\} \\ &= \text{Proj}_U(\mathcal{IR}_{\mathcal{B}} \cap \text{Proj}_V^{-1} \upsilon) \end{aligned} \tag{3.5}$$

and

$$\begin{aligned} \mathcal{IR}_{\mathcal{B}}(\phi) &= \{\upsilon \in \mathbb{R}^{k-p} : \theta = (\phi, \upsilon) \in \mathcal{IR}_{\mathcal{B}}\} \\ &= \text{Proj}_V(\mathcal{IR}_{\mathcal{B}} \cap \text{Proj}_U^{-1} \phi). \end{aligned} \tag{3.6}$$

We observe that $\mathcal{IR}_{\mathcal{B}}(\upsilon)$ and $\mathcal{IR}_{\mathcal{B}}(\phi)$ are also called "fibers" of $\mathcal{IR}_{\mathcal{B}}$ and are associated with fixing a particular $\upsilon \in \Upsilon$ or $\phi \in \Phi$, respectively. We make the following observations.

Observation 3.1 *The sets* $\text{Proj}_U \mathcal{IR}_{\mathcal{B}}$ *and* $\text{Proj}_V \mathcal{IR}_{\mathcal{B}}$ *are semi-algebraic since the projection of a semi-algebraic set is semi-algebraic.*

Observation 3.2 *For each* $\upsilon \in \text{Proj}_V \mathcal{IR}_{\mathcal{B}}$*, the set* $\mathcal{IR}_{\mathcal{B}}(\upsilon)$ *is semi-algebraic since the restriction and projection of a semi-algebraic set are semi-algebraic.*

Observation 3.3 *For each* $\phi \in \text{Proj}_U \mathcal{IR}_{\mathcal{B}}$*, the set* $\mathcal{IR}_{\mathcal{B}}(\phi)$ *is polyhedral since the dependence on* υ *is linear.*

Having defined the subvectors ϕ and υ, recognize that since $M(\theta)$ contains no elements of υ we can write $M(\theta)$ as $M(\phi)$ and $G(\theta)$ as $G(\phi)$. Similarly, for any basis $\mathcal{B}$, we can write $C_{\mathcal{B}}(\theta)$ as $C_{\mathcal{B}}(\phi)$. Next consider the vector $q(\theta)$. Recall from our discussion in Section 2.3 that $q(\theta)$ can be represented as $q(\theta) = q + \Delta Q\theta$. Using the notation introduced above, $q(\theta)$ can be written as $q(\phi, \upsilon) = q + \Delta Q_{\bullet U}\phi + \Delta Q_{\bullet V}\upsilon$.

We recall the following properties from real algebraic geometry.

Definition 3.1 (Definition 2.2.5 of [8]) Let $A \in \mathbb{R}^n$ and $B \in \mathbb{R}^m$ be two semi-algebraic sets. A mapping $f : A \to B$ is *semi-algebraic* if its graph is semi-algebraic in $\mathbb{R}^{m+n}$.

Lemma 3.1 (Theorem 2.8.8 of [8]) *Let A be a semi-algebraic set and* $f : A \to \mathbb{R}^n$ *a semi-algebraic mapping. Then* $dim(A) \geq dim(f(A))$*. If* f *is a bijection from A onto* $f(A)$*, then* $dim(A) = dim(f(A))$*.*

The following proposition results from the above lemma.

Proposition 3.1 *The following hold for each $\phi \in \Phi$:*

1. $dim\left(\bigcup_{\upsilon \in \Upsilon} q(\phi, \upsilon)\right) = dim(\Upsilon)$,
2. $dim\left(\left(\bigcup_{\upsilon \in \Upsilon} q(\phi, \upsilon)\right) \cap C_{\mathcal{B}}(\phi)\right) = dim(\mathcal{IR}_{\mathcal{B}}(\phi))$.

Proof Since in this work we assume that $\triangle Q$ is of full column rank, $\triangle Q_{\bullet V}$ must also be of full column rank. As a result, for each fixed $\phi \in \Phi$, $q(\phi, \upsilon)$ is a bijection onto its image as a function of υ. Furthermore, for each fixed $\phi \in \Phi$, $q(\phi, \upsilon)$ is also semi-algebraic since its graph is the set $\{(\upsilon, q') : \upsilon \in \Upsilon, q' = q + \triangle Q_{\bullet U}\phi + \triangle Q_{\bullet V}\upsilon\}$, which is clearly semi-algebraic since this set only adds an additional polynomial equality. Thus, by Lemma 3.1, the above equalities hold. □

Using the sets and notations introduced so far in this section, together with (2.5), for each f.c.b. $\mathcal{B}$, we provide the following alternate forms of the definitions of invariancy regions and invariant domains:

$$\mathcal{IR}_{\mathcal{B}} := \{(\phi, \upsilon) \in \Theta_{\mathcal{B}} : g_{\mathcal{B}} Adj\ (G(\phi)_{\bullet\mathcal{B}})\, q(\phi, \upsilon) \geq 0\} \tag{3.7}$$

$$\begin{aligned} \mathcal{ID}_{\mathcal{B}} &= \bigcup_{\phi \in \Phi}\left(C_{\mathcal{B}}(\phi) \cap \left(\bigcup_{\upsilon \in \Upsilon} q(\phi, \upsilon)\right)\right) \\ &= \{q(\phi, \upsilon) : G(\phi)_{\bullet\mathcal{B}}^{-1} q(\phi, \upsilon) \geq 0,\ (\phi, \upsilon) \in \Theta\}. \end{aligned} \tag{3.8}$$

Recognize that the definition of an invariancy region in (3.7), together with the results of Propositions 2.1 and 2.2 and Lemma 2.1 and the facts that the elements of $M(\phi)$ are affine functions of ϕ and the elements of $q(\phi, \upsilon)$ are affine functions of ϕ and υ, leads to the following observation.

Observation 3.4 *For each f.c.b. $\mathcal{B}$, the defining inequalities of the invariancy region $\mathcal{IR}_{\mathcal{B}}$ are polynomial in ϕ and linear in υ.*

Throughout the remainder of this section, we develop theoretical results that allow us to exploit the property of invariancy regions described in Observation 3.4. Although this property was discovered in the context of invariancy regions, it is of interest to study in general. For this reason, we establish the following property.

Property 3.1 Given a subset S' of Θ, we say S' satisfies the Polynomial-Linear Property if the following two conditions hold:

- S' is defined by a set of inequalities $f_i(\phi, \upsilon) \leq 0, i \in \{1, \ldots, n\}$ for some $n \in \mathbb{N}$.
- For each $i \in \{1, \ldots, n\}$, f_i is polynomial in ϕ and linear in υ.

In the following section, we consider general sets satisfying the Polynomial-Linear Property. We then develop interesting theoretical consequences of this property and use these results to make important conclusions about invariancy regions.

3.2 Exploiting the Algebraic Structure of an Invariancy Region

Perhaps the most important aspect of invariancy regions for us to study is their dimension. In order to establish the existence of a partition of Θ, we must develop necessary and sufficient conditions for an invariancy region to be of dimension k or $k-1$. To do this, however, we must first establish many more properties of invariancy regions or, more generally, subsets of Θ that satisfy the Polynomial-Linear Property. In the following discussion, we use Ω to denote an arbitrary semi-algebraic subset of Θ satisfying the Polynomial-Linear Property.

To begin, we recall two definitions and a proposition from real algebraic geometry.

Definition 3.2 (Definition 2.4.2 of Bochnak et al. [8]) A semi-algebraic subset A of $\mathbb{R}^n$ is *semi-algebraically connected* if, for every pair of semi-algebraic sets, F_1 and F_2 in A, disjoint and satisfying $F_1 \cup F_2 = A$, one has $F_1 = A$ or $F_2 = A$.

Definition 3.3 (Definition 2.5.12 of Bochnak et al. [8]) A semi-algebraic subset A of $\mathbb{R}^n$ is *semi-algebraically path connected* if, for every $x, y \in A$, there exists a continuous semi-algebraic mapping $\psi : [0, 1] \to A$ such that $\psi(0) = x$ and $\psi(1) = y$.

Proposition 3.2 (Proposition 2.8.5 of Bochnak et al. [8])

1. *Let* $A = \bigcup_{i=1}^{n} A_i$ *be a finite union of semi-algebraic sets. Then* $dim(A) = \max(dim(A_1), \ldots, dim(A_n))$.
2. *Let A and B be two semi-algebraic sets. Then* $dim(A \times B) = dim(A) + dim(B)$.

We now introduce new notation that we will use to show that for any $\Omega \subset \Theta$ satisfying the Polynomial-Linear Property, the dimension of Ω can be expressed in terms of the dimensions of certain subsets of Ω that arise due to the decomposition of θ via ϕ and υ. In a similar fashion to the definitions for $\Phi(\upsilon)$ and $\Upsilon(\phi)$, we introduce the sets

$$\begin{aligned}\Omega(\upsilon) &= \{\phi \in \mathbb{R}^p : \theta = (\phi, \upsilon) \in \Omega\} \\ &= \mathrm{Proj}_U(\mathrm{Proj}_V^{-1} \upsilon \cap \Omega)\end{aligned} \tag{3.9}$$

and

$$\begin{aligned}\Omega(\phi) &= \{\upsilon \in \mathbb{R}^{k-p} : \theta = (\phi, \upsilon) \in \Omega\} \\ &= \mathrm{Proj}_V(\mathrm{Proj}_U^{-1} \phi \cap \Omega).\end{aligned} \tag{3.10}$$

We note here that each of the semi-algebraic sets we use to construct a partition of Θ is always given by a finite set of inequalities. Hence, we assume that each semi-algebraic set considered within this section is also given by a finite set of inequalities. We denote

$$\Omega.\mathcal{D} := \text{The set of defining inequalities of } \Omega. \tag{3.11}$$

Then, for each $\mathcal{S} \subseteq \Omega.\mathcal{D}$, we define

$$\Phi^{\Omega,\mathcal{S},\bullet} :=$$

$$\{\phi \in \text{Proj}_U\, \Omega : \text{The set of redundant inequalities of } \Omega(\phi) \text{ is exactly } \Omega.\mathcal{D} \setminus \mathcal{S}\}. \tag{3.12}$$

In other words, for each $\mathcal{S} \subseteq \Omega.\mathcal{D}$, the set $\Phi^{\Omega,\mathcal{S},\bullet}$ contains all ϕ such that the fiber of Ω resulting from fixing ϕ is defined by the inequalities of $\mathcal{S}$ and none of these are redundant. Additionally, for a given $d \in \mathbb{N}$, we define

$$\Phi^{\Omega,\bullet,d} := \{\phi \in \text{Proj}_U\, \Omega : dim(\Omega(\phi)) = d\}. \tag{3.13}$$

In other words, for each $d \in \mathbb{N}$, the set $\Phi^{\Omega,\bullet,d}$ contains all ϕ such that the fiber of Ω resulting from fixing ϕ is d-dimensional. Finally, for each $\mathcal{S} \subseteq \Omega.\mathcal{D}$ and $d \in \mathbb{N}$, we define

$$\Phi^{\Omega,\mathcal{S},d} := \Phi^{\Omega,\mathcal{S},\bullet} \cap \Phi^{\Omega,\bullet,d} \tag{3.14}$$

and

$$\begin{aligned}\Omega^{\mathcal{S},d} &:= \{(\phi, \upsilon) \in \Omega : \phi \in \Phi^{\Omega,\mathcal{S},d}, \upsilon \in \Omega(\phi)\}\\ &= \text{Proj}_U^{-1}\, \Phi^{\Omega,\mathcal{S},d}.\end{aligned} \tag{3.15}$$

Note that $\Phi^{\Omega,\mathcal{S},d}$ is the set of $\phi \in \mathbb{R}^p$ such that $\Omega(\phi)$ is d-dimensional and for which the set of non-redundant inequalities of $\Omega(\phi)$ is $\mathcal{S}$. Also, $\Omega^{\mathcal{S},d}$ is the set of all $\theta \in \Theta$ that can be formed as (ϕ, υ) where ϕ comes from $\Phi^{\Omega,\mathcal{S},d}$.

From (2.4), (3.6), and (3.15), we make the following observation.

Observation 3.5 *For any set $\Omega \subset \Theta$ satisfying the Polynomial-Linear Property, we have* $\Omega = \bigcup_{d\in\mathbb{N},\mathcal{S}\subseteq\Omega.\mathcal{D}} \Omega^{\mathcal{S},d}$.

We now introduce several theoretical results that provide us with a strategy for determining the dimension of an invariancy region and, moreover, establishing necessary and sufficient conditions for an invariancy region to have dimension k or $k-1$.

Proposition 3.3 *Given $\Omega \subset \Theta$ having Property* 3.1, *the set $\Phi^{\Omega,\bullet,d}$ is semi-algebraic for any $d \in \mathbb{N}$.*

Proof For this discussion, we write $\Phi^{\Omega,\bullet,\geq d'} := \bigcup_{d \geq d'} \Phi^{\Omega,\bullet,d}$. We proceed with the proof by showing that the set $\Phi^{\Omega,\bullet,\geq d}$ is semi-algebraic for each $d \in \mathbb{N}$. This clearly implies that $\Phi^{\Omega,\bullet,d}$ is semi-algebraic since $\Phi^{\Omega,\bullet,d} = \Phi^{\Omega,\bullet,\geq d} \setminus \Phi^{\Omega,\bullet,\geq d+1}$ and intersections and complements of semi-algebraic sets are also semi-algebraic.

We now construct a set $\mathcal{X}^{\Omega,\bullet,\geq d}$ such that $\Phi^{\Omega,\bullet,\geq d}$ is the projection of $\mathcal{X}^{\Omega,\bullet,\geq d}$ onto $\mathbb{R}^U$. We then show that $\mathcal{X}^{\Omega,\bullet,\geq d}$ is semi-algebraic. This is enough to show that $\Phi^{\Omega,\bullet,\geq d}$ is semi-algebraic since projections of semi-algebraic sets are also semi-algebraic. We define $\mathcal{X}^{\Omega,\bullet,\geq d}$ to be the set of $(d+2)$-tuples $(\phi, \upsilon^1, \ldots, \upsilon^{d+1})$ that satisfy the following conditions:

1. For each $i \in \{1, \ldots, d+1\}$, $(\phi, \upsilon^i) \in \Omega$.
2. The matrix $K^{\upsilon^1,\ldots,\upsilon^{d+1}}$ is of full rank, where $K^{\upsilon^1,\ldots,\upsilon^{d+1}} \in \mathbb{R}^{(k-p)\times d}$ is defined so that for each $i \in \{1, \ldots, d\}$, $K^{\upsilon^1,\ldots,\upsilon^{d+1}}_{\bullet i} = \upsilon^i - \upsilon^{d+1}$.

Recognize that condition (1) is enforced by a set of polynomial constraints since Ω is semi-algebraic. Also recognize that condition (2) is satisfied if and only if there exists a $d \times d$ minor of $K^{\upsilon^1,\ldots,\upsilon^{d+1}}$ whose determinant is nonzero. Denote the $d \times d$ minors of $K^{\upsilon^1,\ldots,\upsilon^{d+1}}$ by $D_1, \ldots, D_n$ and note that n must be finite. Then for each $i \in \{1, \ldots, n\}$, consider the set of tuples that satisfy condition (1) and for which D_i is nonzero. Since this set is described by polynomial inequalities, it is a semi-algebraic set. Taking the union of all of these sets over i shows that $\mathcal{X}^{\Omega,\bullet,\geq d}$ is also semi-algebraic as it is a finite union of semi-algebraic sets.

Now, we observe that $(\phi, \upsilon^1, \ldots, \upsilon^{d+1}) \in \mathcal{X}^{\Omega,\bullet,\geq d}$ if and only if $\phi \in \operatorname{Proj}_U \Omega$ and $\upsilon^1, \ldots, \upsilon^{d+1}$ are all in $\Omega(\phi)$. Due to the affine independence of these vectors, it follows that $\Omega(\phi)$ contains a d-simplex and is at least d-dimensional. □

Proposition 3.4 *Given $\Omega \subset \Theta$ satisfying the Polynomial-Linear Property, the set $\Phi^{\Omega,\mathcal{S},\bullet}$ is semi-algebraic for any $\mathcal{S} \subseteq \Omega.\mathcal{D}$.*

Proof For any set of inequalities $\mathcal{T}$ on υ, let $\Psi^{\Omega,\mathcal{T}} = \{(\phi, \upsilon) \in \Theta : \phi \in \operatorname{Proj}_U \Omega$ and υ satisfies $\mathcal{T}\}$. Note that $\Phi^{\Omega,\mathcal{S},\bullet} \subset \operatorname{Proj}_U \Psi^{\Omega,\mathcal{S}}$ and, moreover, $\operatorname{Proj}_U \Psi^{\Omega,\mathcal{S}}$ is semi-algebraic since it is a projection of a semi-algebraic set. We now proceed to construct $\Phi^{\Omega,\mathcal{S},\bullet}$ in two steps. First, we construct the subset of $\operatorname{Proj}_U \Psi^{\Omega,\mathcal{S}}$ containing only ϕ at which $\mathcal{S}$ is not redundant in $\Omega(\phi)$. We then remove from this subset all ϕ at which some $t \in \Omega.\mathcal{D} \setminus \mathcal{S}$ is not redundant.

Given $s \in \mathcal{S}$, consider the set $\operatorname{Proj}_U(\Psi^{\Omega,\mathcal{S}\setminus\{s\}} \setminus \Psi^{\Omega,\mathcal{S}}) \cap \operatorname{Proj}_U \Psi^{\Omega,\mathcal{S}}$. Clearly, this set must contain any ϕ at which s is not redundant in $\Omega(\phi)$ as it is precisely the set of ϕ at which the feasible set without s is strictly larger than the feasible set with s. Additionally, note that this set is semi-algebraic as the projection, intersection, and difference of semi-algebraic sets are semi-algebraic. We then have that the subset of $\operatorname{Proj}_U \Psi^{\Omega,\mathcal{S}}$ containing only ϕ at which $\mathcal{S}$ is not redundant in $\Omega(\phi)$ is given by $\bigcap_{s \in \mathcal{S}} \left(\operatorname{Proj}_U(\Psi^{\Omega,\mathcal{S}\setminus\{s\}} \setminus \Psi^{\Omega,\mathcal{S}}) \cap \operatorname{Proj}_U \Psi^{\Omega,\mathcal{S}}\right)$.

We now remove those ϕ where $\Omega.\mathcal{D} \setminus \mathcal{S}$ is not redundant. For each $t \in \Omega.\mathcal{D} \setminus \mathcal{S}$, consider the set $\operatorname{Proj}_U(\Psi^{\Omega,\mathcal{S}} \setminus \Psi^{\Omega,\mathcal{S}\cup\{t\}})$. Clearly, this set must contain any ϕ at which t is not redundant in $\Omega(\phi)$ as it is precisely the set of ϕ at which the feasible set implied by $\mathcal{S} \cup t$ is smaller than the feasible set implied by $\mathcal{S}$ alone. Again, this set is semi-algebraic as the projection, intersection, and difference of semi-algebraic sets are semi-algebraic. From the work done so far, we now have that

$$\Phi^{\Omega,\mathcal{S},\bullet} = \bigcap_{s\in\mathcal{S}} \left(\text{Proj}_U(\Psi^{\Omega,\mathcal{S}\setminus\{s\}} \setminus \Psi^{\Omega,\mathcal{S}}) \cap \text{Proj}_U\, \Psi^{\Omega,\mathcal{S}}\right) \setminus \left(\bigcup_{t\in\Omega.\mathcal{D}\setminus\mathcal{S}} \text{Proj}_U(\Psi^{\Omega,\mathcal{S}} \setminus \Psi^{\Omega,\mathcal{S}\cup\{t\}})\right)$$

and, moreover, this set is semi-algebraic since the projection, intersection, and difference of semi-algebraic sets are semi-algebraic and because the set difference of a union can be equivalently computed as a sequence of set differences. □

Using Propositions 3.3 and 3.4, the following corollary immediately follows.

Corollary 3.1 *If $\Omega \subset \Theta$ satisfies the Polynomial-Linear Property, the set $\Phi^{\Omega,\mathcal{S},d}$ is semi-algebraic for any $\mathcal{S} \subseteq \Omega.\mathcal{D}$ and $d \in \mathbb{N}$.*

Having now determined that all of our sets of interest are semi-algebraic, we use several additional results from algebraic geometry to establish a lemma and three subsequent propositions that specify the dimensions of these sets. We begin with a definition, modified from [45].

Definition 3.4 By taking the convention that the subsets of $\mathbb{R}^n$ that are considered to be closed are precisely those subsets that are algebraic, the family of all algebraic subsets of $\mathbb{R}^n$ forms a topology on $\mathbb{R}^n$ known as the *Zariski topology*.

Proposition 3.5 (Proposition 2.8.2 of Bochnak et al. [8]) *Let $A \subset \mathbb{R}^n$ be a semi-algebraic set. Then $dim(A) = dim(clos(A)) = dim(clos_{Zar}(A))$, where $clos(A)$ and $clos_{Zar}(A)$ denote the closure and the Zariski closure of A, respectively.*

We recall that the closure of a given set S refers to the intersection of all closed sets that contain S. Thus, the Zariski closure of a set S is the intersection of all algebraic sets that contain S.

Proposition 3.6 (Proposition 2.8.10 of Bochnak et al. [8]) *Let $A \subset \mathbb{R}^n$ be a semi-algebraic set, and let x be a point of A. There exists a semi-algebraic neighborhood N_x of x in A, such that, for any other semi-algebraic neighborhood N'_x of x in A contained in N_x, one has $dim(N_x) = dim(N'_x)$.*

Definition 3.5 (Definition 2.8.11 of [8]) Let $A \subset \mathbb{R}^n$ be a semi-algebraic set, and let x be a point of A. The *local dimension of A at x*, denoted $dim(A_x)$, is $dim(N_x)$, where N_x is as in Proposition 3.6.

Proposition 3.7 (Proposition 2.8.12 of Bochnak et al. [8]) *Let A be a semi-algebraic set of dimension d. Then $\{x \in A : dim(A_x) = d\}$ is a nonempty closed semi-algebraic subset of A.*

We now establish our lemma from the results above. Three subsequent propositions then follow immediately from the lemma.

Lemma 3.2 *If* $\Omega \subset \Theta$ *satisfies the Polynomial-Linear Property, then* $dim\left(\Omega^{S,d}\right) = dim\left(\Phi^{\Omega,S,d}\right) + d$.

Proof By Proposition 3.5, the dimension of a semi-algebraic set is the dimension of its Zariski closure. We note that the Zariski closure of a polytope is its affine span. Therefore, it is enough to compute the dimension of the space $\bigcup_{\phi \in \Phi^{\Omega,S,d}} \{\phi\} \times L^{\phi,d}$, where $L^{\phi,d}$ is a d-dimensional affine span that depends continuously on ϕ. By Proposition 3.2, part (1), it follows that it is enough to look at the dimension of this space locally. Since the d-dimensional affine spans depend continuously on ϕ, for any $\phi \in \Phi^{\Omega,S,d}$, there must exist an open semi-algebraic neighborhood N_ϕ of ϕ in $\Phi^{\Omega,S,d}$ such that $\text{Proj}_U^{-1} N_\phi \cap \Omega^{S,d}$ is in bijective correspondence with $N_\phi \times \mathbb{R}^d$. By Lemma 3.1, $dim\left(N_\phi \times \mathbb{R}^d\right) = dim(\text{Proj}_U^{-1} N_\phi \cap \Omega^{S,d})$. Moreover, by Proposition 3.2, part (2), it follows that $dim\left(N_\phi \times \mathbb{R}^d\right) = dim\left(N_\phi\right) + d$. Finally, by Proposition 3.7, we see that by choosing N_ϕ to be sufficiently small, we have that $dim\left(N_\phi\right)$ is the local dimension, and for some ϕ, $dim(N_\phi) = dim(\Phi^{\Omega,S,d})$. □

Since any $\Omega \subset \Theta$ satisfying the Polynomial-Linear Property can be constructed as $\bigcup_{d \in \mathbb{N}, S \subset \Omega.\mathcal{D}} \Phi^{\Omega,S,d}$ and because $d \leq k$, we immediately have the results contained in the following three propositions.

Proposition 3.8 *Given* $\Omega \subset \Theta$ *satisfying the Polynomial-Linear Property, we have* $dim(\Omega) = \max_{d \in \mathbb{N}, S \subset \Omega.\mathcal{D}} \{dim(\Phi^{\Omega,S,d}) + d\}$.

Proposition 3.9 *Let* $\Omega \subset \Theta$ *satisfying the Polynomial-Linear Property be given. Then* Ω *is* k*-dimensional if and only if there exists* $\Phi' \subseteq \text{Proj}_U \Omega$ *such that* $dim(\Phi') = p$ *and* $dim\left(\Omega(\phi)\right) = k - p$ *for all* $\phi \in \Phi'$.

Proposition 3.10 *Let* $\Omega \subset \Theta$ *satisfying the Polynomial-Linear Property be given for which* $dim(\Omega) \neq k$. *Then* $dim(\Omega) = k - 1$ *if and only if there exists* $\Phi' \subseteq \text{Proj}_U \Omega$ *for which one of the following two conditions holds:*

1. $dim(\Phi') = p$ *and for each* $\phi \in \Phi'$, $dim\left(\Omega(\phi)\right) = (k - p) - 1$.
2. $dim(\Phi') = p - 1$ *and for each* $\phi \in \Phi'$, $dim\left(\Omega(\phi)\right) = k - p$.

We note that Propositions 3.9 and 3.10 provide the necessary conditions to determine the dimension of a given invariancy region. Using these results, we are now able to make several additional observations about invariancy regions. These observations are contained in the following section.

3.3 An Initial Strategy for Partitioning the Parameter Space

In this section, we provide two corollaries that result directly from the theory developed in Section 3.2. In turn, the results of these corollaries give rise to a subsequent proposition that contains the primary theoretical result that ensures the correctness of the methodology we introduce in Chapters 4 and 5. Consider the following two corollaries—the first follows from Proposition 3.9 and the second follows from Proposition 3.10.

Corollary 3.2 *Given a f.c.b.* $\mathcal{B}$, *the invariancy region* $\mathcal{IR}_{\mathcal{B}}$ *is* k*-dimensional if and only if there exists* $\Phi' \subseteq \text{Proj}_U \mathcal{IR}_{\mathcal{B}}$ *such that* $dim(\Phi') = p$ *and* $dim\left(\bigcup_{\upsilon\in\Upsilon} q(\phi,\upsilon)\cap C_{\mathcal{B}}(\phi)\right) = k - p$ *for all* $\phi \in \Phi'$.

Corollary 3.3 *Let a f.c.b.* $\mathcal{B}$ *be given for which* $dim(\mathcal{IR}_{\mathcal{B}}) \neq k$. *Then* $dim(\mathcal{IR}_{\mathcal{B}}) = k - 1$ *if and only if there exists* $\Phi' \subseteq \text{Proj}_U \mathcal{IR}_{\mathcal{B}}$ *for which one of the following two conditions holds:*

1. $dim(\Phi') = p$ *and for each* $\phi \in \Phi'$, $dim\left(\bigcup_{\upsilon\in\Upsilon} q(\phi,\upsilon)\cap C_{\mathcal{B}}(\phi)\right) = (k-p)-1.$
2. $dim(\Phi') = p - 1$ *and for each* $\phi \in \Phi'$, $dim\left(\bigcup_{\upsilon\in\Upsilon} q(\phi,\upsilon)\cap C_{\mathcal{B}}(\phi)\right) = k - p.$

We now use the results of these corollaries and propose another result that will be extremely useful in developing an initial strategy for partitioning Θ. Consider the following proposition.

Proposition 3.11 *Let two f.c.b.'s* $\mathcal{B}_i$ *and* $\mathcal{B}_j$ *be given such that* $\mathcal{IR}_{\mathcal{B}_i}$ *and* $\mathcal{IR}_{\mathcal{B}_j}$ *are each full dimensional and adjacent, i.e.,* $dim\left(\mathcal{IR}_{\mathcal{B}_i}\cap\mathcal{IR}_{\mathcal{B}_j}\right) = k-1$. *Then there exists a sequence of bases* $\left\{\mathcal{B}_n\right\}_{n=i+1}^{j-1}$ *and* $\Phi' \subseteq \cap_{n=i}^{j} \text{Proj}_U \mathcal{IR}_{\mathcal{B}_n}$ *such that: (i)* $dim(\Phi') \geq p-1$, *(ii) for each* $\phi \in \Phi'$, $C_{\mathcal{B}_\gamma}(\phi)$ *and* $C_{\mathcal{B}_{\gamma+1}}(\phi)$ *are adjacent for all* $\gamma \in \{i,\ldots,j-1\}$, *and (iii)* $dim\left(\mathcal{IR}_{\mathcal{B}_\gamma}\right) \geq k-1$ *for all* $\gamma \in \{i+1,\ldots,j-1\}$.

Proof Because $dim\left(\mathcal{IR}_{\mathcal{B}_i}\cap\mathcal{IR}_{\mathcal{B}_j}\right) = k-1$, $dim\left(\text{Proj}_U \mathcal{IR}_{\mathcal{B}_i}\cap\text{Proj}_U \mathcal{IR}_{\mathcal{B}_j}\right)$ is either p or $p-1$. We consider these cases one at a time.

First suppose that $dim\left(\text{Proj}_U \mathcal{IR}_{\mathcal{B}_i}\cap\text{Proj}_U \mathcal{IR}_{\mathcal{B}_j}\right) = p$. Now fix any $\phi' \in \left(\text{Proj}_U \mathcal{IR}_{\mathcal{B}_i}\cap\text{Proj}_U \mathcal{IR}_{\mathcal{B}_j}\right)$ and consider the invariancy regions $\mathcal{IR}_{\mathcal{B}_i}(\phi')$ and $\mathcal{IR}_{\mathcal{B}_j}(\phi')$. Recognize that with ϕ' fixed, $M(\phi')$ is a real valued matrix, and for any basis $\mathcal{B}$, the cone $C_{\mathcal{B}}(\phi')$ is simply the conic combination of vectors with real components. Hence, with ϕ' fixed, we can consider $\mathcal{IR}_{\mathcal{B}_i}(\phi')$ and $\mathcal{IR}_{\mathcal{B}_j}(\phi')$ in the context of the works of [11] and [2], since in these works M is a real valued matrix. Then by Theorem 5.10 of [11], there exists a sequence of invariancy regions $\{\mathcal{IR}_{\mathcal{B}_n}(\phi')\}_{n=i+1}^{j-1}$, and by extension, a sequence of bases $\{B_n\}_{n=i+1}^{j-1}$, such that $\mathcal{C}_{\mathcal{B}_\gamma}(\phi')$ and $\mathcal{C}_{\mathcal{B}_{\gamma+1}}(\phi')$ are adjacent for all $\gamma \in \{i, \ldots, j-1\}$ and $dim\left(\bigcup_{\upsilon\in\Upsilon} q(\phi', \upsilon) \cap \mathcal{C}_{\mathcal{B}_\gamma}(\phi')\right) \geq (k-p)-1$ for all $\gamma \in \{i+1, \ldots, j-1\}$. We say that such a sequence of bases is *valid* for ϕ'. Recognize that valid sequences are not necessarily unique, and furthermore, the same sequence may not be valid for distinct $\phi^*, \phi^{**} \in \left(\text{Proj}_U\, \mathcal{IR}_{\mathcal{B}_i} \cap \text{Proj}_U\, \mathcal{IR}_{\mathcal{B}_j}\right)$.

For a given $\phi \in \left(\text{Proj}_U\, \mathcal{IR}_{\mathcal{B}_i} \cap \text{Proj}_U\, \mathcal{IR}_{\mathcal{B}_j}\right)$, let $\mathscr{C}(\phi)$ represent the set of all valid sequences for ϕ and define $\mathscr{C} := \cup_{\phi\in\left(\text{Proj}_U\, \mathcal{IR}_{\mathcal{B}_i}\cap\text{Proj}_U\, \mathcal{IR}_{\mathcal{B}_j}\right)} \mathscr{C}(\phi)$. Then for each $\mathscr{S} \in \mathscr{C}$ define the set

$$\mathscr{V}(\mathscr{S}) := \left\{\phi \in \left(\text{Proj}_U\, \mathcal{IR}_{\mathcal{B}_i} \cap \text{Proj}_U\, \mathcal{IR}_{\mathcal{B}_j}\right) : \text{ sequence } \mathscr{S} \text{ is valid for } \phi\right\}. \tag{3.16}$$

Recognize that the following are true:

1. $|\mathscr{C}| < \infty$.
2. $\bigcup_{\mathscr{S}\in\mathscr{C}} \mathscr{V}(\mathscr{S}) = \left(\text{Proj}_U\, \mathcal{IR}_{\mathcal{B}_i} \cap \text{Proj}_U\, \mathcal{IR}_{\mathcal{B}_j}\right)$.
3. For each $\mathscr{S} \in \mathscr{C}$, the set $\mathscr{V}(\mathscr{S})$ is semi-algebraic.

The first is due to the fact that there are a finite number of bases and hence a finite number of sequences. The second is obvious. The third is due to the fact that $\mathscr{V}(\mathscr{S})$ can be represented as

$$\begin{aligned}\mathscr{V}(\mathscr{S}) &= \left(\bigcap_{\mathcal{B}\in\mathscr{S}} \Phi^{\mathcal{IR}_{\mathcal{B}},\bullet,(\geq k-p-1)}\right) \cap \left(\bigcap_{(\mathcal{B}_i,\mathcal{B}_j)\in\mathscr{S}} \left\{\phi : \mathcal{C}_{\mathcal{B}_i}(\phi) \text{ is adjacent to } \mathcal{C}_{\mathcal{B}_j}(\phi)\right\}\right)\\ &= \left(\bigcap_{\mathcal{B}\in\mathscr{S}} \Phi^{\mathcal{IR}_{\mathcal{B}},\bullet,(\geq k-p-1)}\right) \cap \left(\bigcap_{(\mathcal{B}_i,\mathcal{B}_j)\in\mathscr{S}} \left\{\phi : rank\left(G_{\bullet(\mathcal{B}_i\cap\mathcal{B}_j)}(\phi)\right) = h-1\right\}\right),\end{aligned}$$

where we have used a similar notation as in the proof of Proposition 3.3 to denote the set $\{\phi \in \text{Proj}_U\, \mathcal{IR}_{\mathcal{B}} : dim(\mathcal{IR}_{\mathcal{B}}(\phi)) \geq d\}$ as $\Phi^{\mathcal{IR}_{\mathcal{B}},\bullet,\geq d}$. We now argue that the above set is semi-algebraic by showing the following: (i) $\Phi^{\mathcal{IR}_{\mathcal{B}},\bullet,(\geq k-p-1)}$ is semi-algebraic for each $\mathcal{B} \in \mathscr{S}$ and (ii) the set $\left\{\phi : rank\left(G_{\bullet(\mathcal{B}_i\cap\mathcal{B}_j)}(\phi)\right) = h-1\right\}$ is semi-algebraic for each pair $(\mathcal{B}_i, \mathcal{B}_j) \in \mathscr{S}$. The former is clear from Proposition 3.3 and the fact that invariancy regions satisfy the Polynomial-Linear Property. The arguments needed to show that the latter are analogous to those used in the proof of Proposition 3.3 in which we showed that conditions on the rank of a matrix can be imposed using a set of polynomial inequalities. Now, since (1), (2), and (3) hold, we have by Proposition 3.2 that there exists some $\mathscr{S}' \in \mathscr{C}$ for which

$dim(\mathscr{V}(\mathscr{S}')) = p$. Hence, if we let $\Phi' = \mathscr{V}(\mathscr{S}')$, together $\mathscr{S}'$ and Φ' satisfy conditions (i) and (ii) of the proposition. Furthermore, since $dim(\Phi') = p$ and $dim\left(\bigcup_{\upsilon\in\Upsilon} q(\phi,\upsilon)\cap \mathcal{C}_{\mathcal{B}}(\phi)\right) \geq (k-p)-1$ for all $\phi \in \Phi'$ and $\mathcal{B} \in \mathscr{S}$, condition (iii) of the proposition is also satisfied by Corollary 3.3.

Now consider the case in which $dim\left(\text{Proj}_U\, \mathcal{IR}_{\mathcal{B}_i} \cap \text{Proj}_U\, \mathcal{IR}_{\mathcal{B}_j}\right) = p-1$. Recognize that $\left(\text{Proj}_U\, \mathcal{IR}_{\mathcal{B}_i} \cap \text{Proj}_U\, \mathcal{IR}_{\mathcal{B}_j}\right)$ satisfies the Polynomial-Linear Property, and therefore by Proposition 3.10, we have that since $dim\left(\mathcal{IR}_{\mathcal{B}_i} \cap \mathcal{IR}_{\mathcal{B}_j}\right) = k-1$, there must exist a $(p-1)$-dimensional subset Φ'' of $\left(\text{Proj}_U\, \mathcal{IR}_{\mathcal{B}_i} \cap \text{Proj}_U\, \mathcal{IR}_{\mathcal{B}_j}\right)$ such that $dim\left(\bigcup_{\upsilon\in\Upsilon} q(\phi,\upsilon)\cap \mathcal{C}_{\mathcal{B}_i}(\phi)\right) = k-p$ for all $\phi \in \Phi''$ and $dim\left(\bigcup_{\upsilon\in\Upsilon} q(\phi,\upsilon)\cap \mathcal{C}_{\mathcal{B}_j}(\phi)\right) = k-p$ for all $\phi \in \Phi''$. Recognize that this can only happen if for each $\phi \in \Phi''$, $\mathcal{C}_{\mathcal{B}_i}(\phi)$ and $\mathcal{C}_{\mathcal{B}_j}(\phi)$ share a facet $\mathscr{F}(\phi)$ for which: (i) $dim\left(\mathscr{F}(\phi)\right) \geq k-p$ and (ii) $dim\left(\mathscr{F}(\phi)\cap\left(\bigcup_{\upsilon\in\Upsilon} q(\phi,\upsilon)\right)\right) = k-p$. Now, recall the following facts: (i) for every f.c.b. $\mathcal{B}$ and every $\phi \in \Phi$, $\mathcal{C}_{\mathcal{B}}(\phi)$ is an h-dimensional cone in $\mathbb{R}^h$ and (ii) because $M(\phi)$ is sufficient for each $\phi \in \Phi$, we have $\mathcal{K}(M(\phi))$ is convex for each $\phi \in \Phi$ and $\cup_{\text{feasible bases } \mathcal{B}}\, \mathcal{C}_{\mathcal{B}}(\phi)$ forms a partition of $\mathcal{K}(M(\phi))$ for each $\phi \in \Phi$. From these facts, recognize that for every $\phi \in \Phi''$, every $\tau \in relint\left(\mathscr{F}(\phi)\cap \bigcup_{\upsilon\in\Upsilon} q(\phi,\upsilon)\right)$, and every $\tau' \in \mathbb{R}^h$, there exists $\epsilon(\tau,\tau',\phi) > 0$ such that for all $\epsilon \in (0, \epsilon(\tau,\tau',\phi)]$, we have either: (i) $\tau+\epsilon\tau' \notin \mathcal{K}(M(\phi))$ or (ii) $\tau+\epsilon\tau' \in \mathcal{K}(M(\phi))$ and the parametric complementary cone containing $\tau+\epsilon\tau'$ also contains $\mathscr{F}(\phi)$. For each $\phi \in \Phi''$, define the set

$$\mathscr{E}(\phi) :=$$

$$\left\{\mathcal{B} : \exists \tau \in relint\left(\mathscr{F}(\phi)\cap \bigcup_{\upsilon\in\Upsilon} q(\phi,\upsilon)\right), \tau' \in \mathbb{R}^h \text{ such that } \tau + \epsilon(\tau,\tau',\phi)\tau' \in \mathcal{C}_{\mathcal{B}}(\phi)\right\}.$$

Further recognize that due to the convexity of $\mathcal{K}(M(\phi))$ for each $\phi \in \Phi$, there must exist a subset $\{\mathcal{B}_1, \ldots, \mathcal{B}_m\}$ of bases in $\mathscr{E}(\phi)$ such that the sequence $\left\{\mathcal{B}_n\right\}_{n=1}^{m}$ satisfies the following properties: (i) $\mathcal{B}_1 = \mathcal{B}_i$, (ii) $\mathcal{B}_m = \mathcal{B}_j$, (iii) $\mathcal{C}_{\mathcal{B}_n}(\phi)$ and $\mathcal{C}_{\mathcal{B}_{n+1}}(\phi)$ are adjacent for all $n \in \{1, \ldots, m-1\}$, and (iv) $\mathscr{F}(\phi) \subset \mathcal{C}_{\mathcal{B}_n}(\phi)$ for all $n \in \{1, \ldots, m\}$. As with the case in which $dim\left(\text{Proj}_U\, \mathcal{IR}_{\mathcal{B}_i} \cap \text{Proj}_U\, \mathcal{IR}_{\mathcal{B}_j}\right) = p$, we say that this sequence is *valid* for ϕ. Hence, as we did in the previous case, for each sequence $\mathscr{S}$ valid for some $\phi \in \Phi''$, we can construct the set $\mathscr{V}(\mathscr{S})$ of $\phi \in \Phi''$ for which $\mathscr{S}$ is valid. Then since we showed that there are a finite number of valid sequences, $\bigcup_{\mathscr{S}\in\mathscr{C}} \mathscr{V}(\mathscr{S}) = \left(\text{Proj}_U\, \mathcal{IR}_{\mathcal{B}_i} \cap \text{Proj}_U\, \mathcal{IR}_{\mathcal{B}_j}\right)$, and for each valid sequence $\mathscr{S}$, the set $\mathscr{V}(\mathscr{S})$ is semi-algebraic, we have from Proposition 3.2 that there must exist a valid sequence $\mathscr{S}'$ such that $dim(\mathscr{V}(\mathscr{S}')) = p - 1$. Let $\Phi' = \mathscr{V}(\mathscr{S}')$. Then clearly Φ' is a $(p - 1)$-dimensional subset of Φ'' such that a single sequence of bases $\mathscr{S}$ is valid for all $\phi \in \Phi'$. Hence, conditions (i) and (ii) of the proposition also hold when $dim\left(\text{Proj}_U\, \mathcal{IR}_{\mathcal{B}_i} \cap \text{Proj}_U\, \mathcal{IR}_{\mathcal{B}_j}\right) = p - 1$.

Furthermore, since $dim(\Phi') = p - 1$ and $dim\left(\bigcup_{\upsilon\in\Upsilon} q(\phi, \upsilon) \cap \mathcal{C}_{\mathcal{B}}(\phi)\right) = k - p$ for all $\phi \in \Phi'$ and $\mathcal{B} \in \mathscr{S}$, condition (iii) of the proposition is also satisfied by Proposition 3.10. □

To aid in visualization of some of the concepts introduced in the proof of Proposition 3.11, particularly the case in which $dim\left(\text{Proj}_U\, \mathcal{IR}_{\mathcal{B}_i} \cap \text{Proj}_U\, \mathcal{IR}_{\mathcal{B}_j}\right) = p-1$, we include Figure 3.1 that displays an example of parametric complementary cones that are not adjacent, but do share a $(k-1)$-dimensional facet. Note that this example is specific to the special case in which $h = 3$ and $p = k - 1$. Observe, particularly from the top view in Figure 3.1b, that because $\mathcal{K}(M(\phi))$ is convex and partitioned by the set of parametric complementary cones, the missing space "between" cones $\mathcal{C}_{\mathcal{B}_i}(\phi)$ and $\mathcal{C}_{\mathcal{B}_j}(\phi)$ must also be partitioned by other parametric complementary

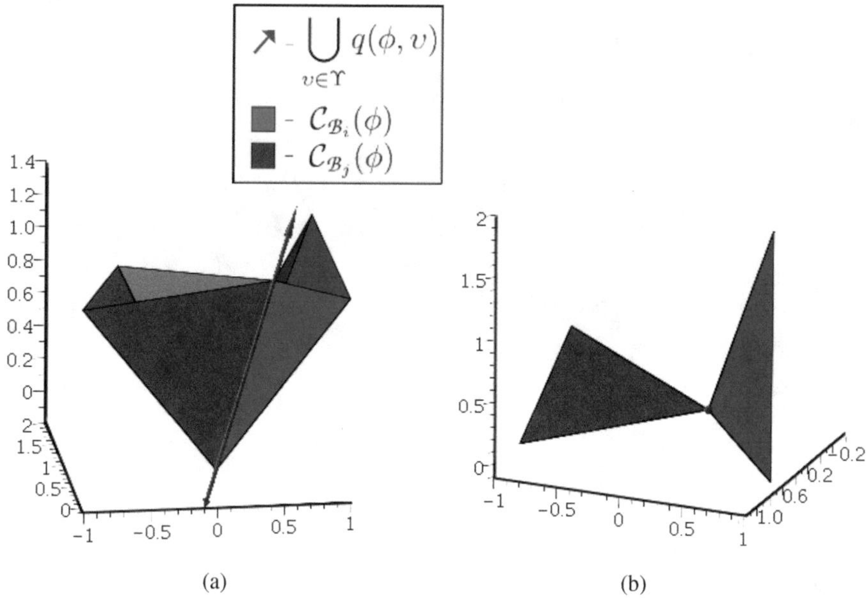

Fig. 3.1 Example of $(k-p)$-dimensional intersection of $\bigcup_{\upsilon\in\Upsilon} q(\phi, \upsilon)$ with two distinct complementary cones. ($h = 3$, $p = k - 1$). (**a**) Side view (**b**) Top view

cones, and furthermore, there must be a subset C of these cones such that for each cone in C, the intersection of the cone with $\bigcup_{\upsilon \in \Upsilon} q(\phi, \upsilon)$ is $(k-1)$-dimensional. Hence, there must be a sequence of cones in C, and by extension a sequence of bases, that satisfy the conditions of Proposition 3.11.

Note that Proposition 3.11 indicates that one strategy for partitioning Θ is to begin with a full dimensional invariancy region, compute sequences of bases that yield each adjacent full dimensional invariancy region, and repeat this process for each discovered region until no undiscovered invariancy regions exist within Θ. In the next chapter we introduce theoretical results that we use to carry out these procedures.

Chapter 4
Phase 2: Partitioning the Parameter Space

In this chapter we introduce an algorithm for partitioning Θ. The presented algorithm is designed to be utilized only after the execution of the phase 1 algorithm (see Chapter 5) and therefore relies on the initial f.c.b. $\mathcal{B}_0$ returned by the phase 1 algorithm. We save details for Chapter 5, but note here that, under Assumptions 1.1 and 1.2, the phase 1 algorithm is guaranteed to return either a f.c.b. whose associated invariancy region is full dimensional or a certificate that no such f.c.b. exists. Since the latter case is not interesting, we assume from here that $dim\left(\mathcal{IR}_{\mathcal{B}_0}\right) = k$. The phase 2 algorithm, an iterative process, then begins by identifying all f.c.b.'s whose associated invariancy regions are adjacent to $\mathcal{IR}_{\mathcal{B}_0}$, i.e., have $(k-1)$-dimensional intersection with $\mathcal{IR}_{\mathcal{B}_0}$. Details on how such f.c.b.'s are obtained are saved for Section 4.2, but we do point out that by Proposition 3.11 we know that all such f.c.b.'s can be discovered by considering only f.c.b.'s whose associated invariancy regions are of dimension $k-1$ or higher. In each subsequent iteration of the phase 2 algorithm, a known f.c.b., say $\mathcal{B}'$, having either a k or $(k-1)$-dimensional invariancy region, is processed by computing all previously undiscovered f.c.b.'s whose associated invariancy regions are adjacent to $\mathcal{IR}_{\mathcal{B}'}$. The algorithm ceases when there are no new k or $(k-1)$-dimensional invariancy regions left to discover. Recognize that whenever Assumption 1.2 is satisfied, $\hat{\Theta}$ must be a single connected set and, therefore, after completion of the aforementioned process, the set of discovered f.c.b.'s having k-dimensional invariancy regions forms the desired partition of Θ. Note that the partition of Θ contains only k-dimensional invariancy regions because any lower dimensional region must be contained within a union of k-dimensional regions that are part of the partition. However, the algorithm we use to construct these k-dimensional regions requires the consideration of $(k-1)$-dimensional regions.

We note here that while the dependence of the strategy proposed above on Assumption 1.2 is clear, the dependence on Assumption 1.1 is not as obvious. It is clear that the success of the process described above depends primarily on one's ability to discover all full dimensional invariancy regions that are adjacent

N. Adelgren, *Advancing Parametric Optimization*, SpringerBriefs in Optimization,
https://doi.org/10.1007/978-3-030-61821-6_4

to a given full dimensional invariancy region. As will be discussed in detail later, discovering adjacent *regions* depends on the discovery of adjacent *bases* and, moreover, discovery of adjacent bases relies on the underlying structure of the (mp)LCP problem as was discussed in Chapter 2. Specifically, bases are considered adjacent if and only if their associated complementary cones are adjacent (see Definition 2.14). Thus, in order to ensure that the discovery of ajacent bases is possible, we require that: (i) for any given $\theta \in \Theta$, all complementary cones are disjoint in their relative interiors, and (ii) for any given $\theta \in \Theta$, the union of all complementary cones is a convex set. As noted at the end of Chapter 2, to ensure that these properties hold, we require that for any given $\theta \in \Theta$, the matrix $M(\theta)$ be both column sufficient and a Q_0 matrix. As the class of sufficient matrices is the largest class of matrices known to contain both column sufficient matrices and Q_0 matrices, this gives rise to Assumption 1.1. Additional insight into the importance of Assumptions 1.1 and 1.2 can be gained by studying the examples we present in Section 6.1.

The high-level idea of the process described in the opening paragraph of this chapter is presented in Algorithm 1. Additional details are provided as the chapter progresses. For use in Algorithm 1, we define for each f.c.b. $\mathcal{B}$ and each $i \in \mathcal{B}$,

$$\hbar^i_{\mathcal{B}} := \left\{(\phi, \upsilon) \in \mathbb{R}^k : (Adj(G(\phi)_{\bullet\mathcal{B}}))_{i\bullet} q(\phi, \upsilon) = 0\right\}, \tag{4.1}$$

the hypersurface in $\mathbb{R}^k$ that implies the defining constraint of $\mathcal{IR}_{\mathcal{B}}$ associated with i. Additionally, for each f.c.b. $\mathcal{B}$, we define the set

$$\mathrm{F}_{\mathcal{B}} := \left\{i \in \mathcal{B} : dim(\hbar^i_{\mathcal{B}} \cap \mathcal{IR}_{\mathcal{B}}) = k - 1\right\} \tag{4.2}$$

of indices in $\mathcal{B}$ whose associated hypersurfaces form $(k-1)$-dimensional boundaries of $\mathcal{IR}_{\mathcal{B}}$. Finally, to aid in efficient computation, we define the set

$$\mathscr{B} := \{\text{f.c.b.'s } \mathcal{B}' : \mathcal{B}' \text{ has been discovered and processed}\}. \tag{4.3}$$

By dynamically updating the set $\mathscr{B}$ throughout the partitioning procedure, we are able to avoid unnecessary computation by ensuring that no f.c.b. is considered more than once.

In Algorithm 1, $\mathcal{S}$ represents a stack of f.c.b.'s whose associated invariancy regions are k-dimensional. Each f.c.b. $\mathcal{B}$ on the stack is processed on lines 3–6. First, on line 3 we build the set $\mathrm{F}_{\mathcal{B}}$ using the subroutine BUILDF (Algorithm 2). Then on lines 4–6 we use the subroutine GETADJACENTREGIONSACROSS (Algorithm 4) to obtain the set $\mathcal{S}'$ of f.c.b.'s whose associated invariancy regions are k-dimensional and adjacent to $\mathcal{IR}_{\mathcal{B}}$ across $\hbar^i_{\mathcal{B}}$, update the stack $\mathcal{S}$, and update $\mathcal{P}$, which upon termination is a set of invariancy regions that partition $\hat{\Theta}$. We note that a full

Algorithm 1 PARTITION$\Theta(\mathcal{B}_0)$—Partition the parameter space Θ

Input: An initial f.c.b. $\mathcal{B}_0$ such that $dim(\mathcal{IR}_{\mathcal{B}_0}) = k$.
Output: A partition of $\hat{\Theta}$, denoted $\mathcal{P}$.

```
Let S = {B_0} and P = {IR_{B_0}}.
while S ≠ ∅ do select B from S.
    F_B = BUILDF(B)
    for i ∈ F_B do
        Let (S', 𝒷) = GETADJACENTREGIONSACROSS(B, i, 𝒷) and set S = S ∪ S'.
        for B' ∈ S' do set P = P ∪ IR_{B'}.
Return P.
```

partition of Θ is then available as the set of invariancy regions in $\mathcal{P}$ together with the infeasible subset of Θ, which is given by $\Theta \setminus \bigcup\limits_{S \in \mathcal{P}} S$.

We note here that because phase 1 of the procedure presented in this work is guaranteed to return a full dimensional invariancy region (discussed in detail in Chapter 5), for much of our discussion in the current chapter it is safe for us to assume that, given a f.c.b. $\mathcal{B}$, we have $dim(\mathcal{IR}_{\mathcal{B}}) = k$. Hence, we work under this assumption until we reach Section 4.2.4.

We now consider the subroutine BUILDF. For any f.c.b. $\mathcal{B}$ and index $i \in \mathcal{B}$, we define the following sets that will be useful during the remainder of this discussion:

$$\mathrm{Z}_{\mathcal{B}} := \left\{ j \in \mathcal{B} : \left(Adj\left(G(\phi)_{\bullet\mathcal{B}}\right)\right)_{j\bullet} q(\phi, \upsilon) = 0 \quad \forall \phi \in \Phi, \upsilon \in \Upsilon \right\} \tag{4.4}$$

$$\mathrm{E}_{\mathcal{B}} := \left\{ j \in \mathcal{B} : \hbar^{j}_{\mathcal{B}} \cap \mathcal{IR}_{\mathcal{B}} = \emptyset \right\} \tag{4.5}$$

$$\mathrm{H}^{i}_{\mathcal{B}} := \left\{ j \in \mathcal{B} \setminus (\mathrm{E}_{\mathcal{B}} \cup \{i\}) : \left(\hbar^{i}_{\mathcal{B}} \cap \mathcal{IR}_{\mathcal{B}}\right) \subseteq \left(\hbar^{j}_{\mathcal{B}} \cap \mathcal{IR}_{\mathcal{B}}\right) \right\}. \tag{4.6}$$

Here $\mathrm{Z}_{\mathcal{B}}$ is the set of indices in $\mathcal{B}$ for which the right hand side (RHS) of the augmented matrix $\left[\, G(\phi)^{-1}_{\bullet\mathcal{B}} G(\phi) \,\middle|\, G(\phi)^{-1}_{\bullet\mathcal{B}} q(\phi, \upsilon) \,\right]$ is identically zero (in the remainder of this work, we use the symbol "$\equiv$" to denote the property of being *identically equal*). For given $\phi \in \Phi$ and $\upsilon \in \Upsilon$, these RHS values can be interpreted as the multipliers on the columns of $G(\phi)_{\bullet\mathcal{B}}$ needed to represent $q(\phi, \upsilon)$ as a linear combination of the columns of $G(\phi)$. Thus, if the RHS value is identically zero for some index i, this indicates that the column $G(\phi)_{\bullet i}$ is unnecessary in the representation of $q(\phi, \upsilon)$. There is also another interpretation. Notice from (2.4) that for each $i \in \mathcal{B}$, $\left(G(\phi)^{-1}_{\bullet\mathcal{B}}\right)_{i\bullet} q(\phi, \upsilon) \geq 0$ is a defining inequality of $\mathcal{IR}_{\mathcal{B}}$. Thus, if there is some $i \in \mathcal{B}$ for which the RHS is identically zero, the associated defining inequality of $\mathcal{IR}_{\mathcal{B}}$ is $0 \geq 0$, which is trivially satisfied. The set $\mathrm{E}_{\mathcal{B}}$ is then the set of indices $j \in \mathcal{B}$ for which $\hbar^{j}_{\mathcal{B}}$ does not intersect $\mathcal{IR}_{\mathcal{B}}$. Finally, consider $\mathrm{H}^{i}_{\mathcal{B}}$. Given an index $i \in \mathcal{B}$, the set $\mathrm{H}^{i}_{\mathcal{B}}$ is the set of indices j in $\mathcal{B} \setminus (\mathrm{E}_{\mathcal{B}} \cup \{i\})$ such that the intersection of $\hbar^{i}_{\mathcal{B}}$ and $\mathcal{IR}_{\mathcal{B}}$ is a subset of the intersection of $\hbar^{j}_{\mathcal{B}}$ and $\mathcal{IR}_{\mathcal{B}}$.

Recognize that given $i \in \mathcal{B}$, for each $j \in \mathrm{H}^i_{\mathcal{B}} \subseteq \mathcal{B} \setminus (\mathrm{E}_{\mathcal{B}} \cup \{i\})$, every point in $\mathcal{IR}_{\mathcal{B}}$ that satisfies the defining constraint of $\mathcal{IR}_{\mathcal{B}}$ associated with i at equality also satisfies the defining constraints of $\mathcal{IR}_{\mathcal{B}}$ associated with j at equality.

In Section 4.1 we establish that each time a f.c.b. $\mathcal{B}$ is considered throughout the presented procedure for partitioning Θ, the matrix $\left[\, G(\phi)^{-1}_{\bullet\mathcal{B}} G(\phi) \,\vdots\, G(\phi)^{-1}_{\bullet\mathcal{B}} q(\phi, \upsilon) \,\right]$ is always readily available. Hence, it is clear from (4.4) that $\mathrm{Z}_{\mathcal{B}}$ can be constructed easily by observing the RHS of this matrix. We save the discussion on constructing $\mathrm{E}_{\mathcal{B}}$ and $\mathrm{H}^i_{\mathcal{B}}$ until Section 4.2.1.

Note that for the remainder of this chapter, when a f.c.b. $\mathcal{B}$ is passed to any routine or subroutine (such as BUILDF), we assume that $\mathrm{Z}_{\mathcal{B}}$, $\mathrm{E}_{\mathcal{B}}$, and $\mathrm{H}^i_{\mathcal{B}}$ have been constructed and are passed into the routine along with $\mathcal{B}$.

Consider the following observation and a proposition addressing whether or not for a given f.c.b. $\mathcal{B}$ and $i \in \mathcal{B}$, the hypersurface $\hbar^i_{\mathcal{B}}$ forms a $(k-1)$-dimensional boundary of $\mathcal{IR}_{\mathcal{B}}$.

Observation 4.1 *Let a f.c.b. $\mathcal{B}$ and an index $i \in \mathcal{B}$ be given. It is clear from* (4.4) *and* (4.5) *that if $i \in (Z_{\mathcal{B}} \cup E_{\mathcal{B}})$, then $\hbar^i_{\mathcal{B}}$ does not form a $(k-1)$-dimensional boundary of $\mathcal{IR}_{\mathcal{B}}$.*

Proposition 4.1 *For a given f.c.b. $\mathcal{B}$, if there exist $i \in \left(\mathcal{B} \setminus (Z_{\mathcal{B}} \cup E_{\mathcal{B}})\right)$, $\phi' \in \Phi$, and $\upsilon' \in \Upsilon$ such that $g_{\mathcal{B}}\left(Adj(G(\phi')_{\bullet\mathcal{B}})\right)_{j\bullet} q(\phi', \upsilon') > 0$ for all $j \in \left(\mathcal{B} \setminus \left(Z_{\mathcal{B}} \cup H^i_{\mathcal{B}} \cup \{i\}\right)\right)$ and $\left(Adj(G(\phi')_{\bullet\mathcal{B}})\right)_{i\bullet} q(\phi', \upsilon') = 0$, then $\hbar^i_{\mathcal{B}}$ forms a $(k-1)$-dimensional boundary of $\mathcal{IR}_{\mathcal{B}}$. Hence, if the following NLP has a strictly positive optimal value, $\hbar^i_{\mathcal{B}}$ forms a $(k-1)$-dimensional boundary of $\mathcal{IR}_{\mathcal{B}}$.*

$$
\begin{aligned}
NLP_F(\mathcal{B}, i) := \max_{\lambda,\phi,\upsilon} \quad & \lambda \\
\text{s.t.} \quad & g_{\mathcal{B}}\left(Adj(G(\phi)_{\bullet\mathcal{B}})\right)_{j\bullet} q(\phi, \upsilon) \geq \lambda \quad \forall j \in \left(\mathcal{B} \setminus \left(Z_{\mathcal{B}} \cup H^i_{\mathcal{B}} \cup \{i\}\right)\right) \\
& \left(Adj(G(\phi)_{\bullet\mathcal{B}})\right)_{i\bullet} q(\phi, \upsilon) = 0 \\
& \phi \in \Phi, \ \upsilon \in \Upsilon
\end{aligned}
\tag{4.7}
$$

Proof Suppose $(\lambda^*, \phi^*, \upsilon^*)$ is a solution to $NLP_F(\mathcal{B}, i)$ and $\lambda^* > 0$. Notice that since $(\lambda^*, \phi^*, \upsilon^*)$ is feasible to $NLP_F(\mathcal{B}, i)$, we have $(Adj(G(\phi^*)_{\bullet\mathcal{B}}))_{i\bullet} q(\phi^*, \upsilon^*) = 0$, which clearly shows that $\theta^* = (\phi^*, \upsilon^*)$ lies on the hypersurface $\hbar^i_{\mathcal{B}}$. We proceed by showing that $\theta^* \in \mathcal{IR}_{\mathcal{B}}$ and that the defining constraints of $\mathcal{IR}_{\mathcal{B}}$ that are implied by $\hbar^i_{\mathcal{B}}$ cannot be removed without adding new points to $\mathcal{IR}_{\mathcal{B}}$, i.e., $\hbar^i_{\mathcal{B}}$ forms a $(k-1)$-dimensional boundary of $\mathcal{IR}_{\mathcal{B}}$.

We now show that $\theta^* \in \mathcal{IR}_{\mathcal{B}}$ by showing that θ^* satisfies all defining constraints of $\mathcal{IR}_{\mathcal{B}}$. Recognize that since θ^* satisfies the constraints of $NLP_F(\mathcal{B}, i)$, we have that θ^* satisfies the defining constraints of $\mathcal{IR}_{\mathcal{B}}$ associated with each element of $\mathcal{B} \setminus \left(\mathrm{Z}_{\mathcal{B}} \cup \mathrm{H}^i_{\mathcal{B}}\right)$. From (4.6), notice that for each $\ell \in \mathcal{B}$, if $\ell \in \mathrm{H}^i_{\mathcal{B}}$, then $(\hbar^i_{\mathcal{B}} \cap \mathcal{IR}_{\mathcal{B}}) \subseteq (\hbar^\ell_{\mathcal{B}} \cap \mathcal{IR}_{\mathcal{B}})$. Hence, since $(Adj(G(\phi^*)_{\bullet\mathcal{B}}))_{i\bullet} q(\phi^*, \upsilon^*) = 0$, it is clear from (4.1) that $(Adj(G(\phi^*)_{\bullet\mathcal{B}}))_{\ell\bullet} q(\phi^*, \upsilon^*) = 0$ for all $\ell \in \mathrm{H}^i_{\mathcal{B}}$. This also shows that θ^* satisfies all defining constraints of $\mathcal{IR}_{\mathcal{B}}$ associated with elements of $\mathrm{H}^i_{\mathcal{B}}$. Next, recall from (4.4) that for all $j \in \mathrm{Z}_{\mathcal{B}}$, $(Adj(G(\phi)_{\bullet\mathcal{B}}))_{j\bullet}q(\phi, \upsilon) = 0$ for all $\phi \in \Phi$ and $\upsilon \in \Upsilon$. Therefore, θ^* trivially satisfies all defining constraints of $\mathcal{IR}_{\mathcal{B}}$ associated with elements of $\mathrm{Z}_{\mathcal{B}}$. Thus, we now have that θ^* satisfies all defining constraints of $\mathcal{IR}_{\mathcal{B}}$. Furthermore, since $\lambda^* > 0$, there must exist $\epsilon > 0$ such that all $\theta = (\phi, \upsilon)$ in the ball of radius ϵ centered at θ^* satisfy $(Adj(G(\phi)_{\bullet\mathcal{B}}))_{j\bullet}q(\phi, \upsilon) \geq 0$ for all $j \in \left(\mathcal{B} \setminus \left(\mathrm{H}^i_{\mathcal{B}} \cup \{i\}\right)\right)$. Hence, since θ^* lies on $\hbar^i_{\mathcal{B}}$, this ϵ-ball must contain a $\hat{\theta}$ satisfying all the defining inequalities of $\mathcal{IR}_{\mathcal{B}}$ except those implied by $\hbar^i_{\mathcal{B}}$. Thus, the inequalities implied by $\hbar^i_{\mathcal{B}}$ cannot be removed from the description of $\mathcal{IR}_{\mathcal{B}}$ without altering its structure, and therefore $\hbar^i_{\mathcal{B}}$ forms a $(k-1)$-dimensional boundary of $\mathcal{IR}_{\mathcal{B}}$. □

We note here that due to the fact that Θ is assumed to be a bounded semi-algebraic set, NLP_F will always have a bounded feasible region. Moreover, Θ being bounded guarantees that every NLP we introduce in this chapter will also have a bounded feasible region.

The following observations result from Proposition 4.1.

Observation 4.2 *The converse of Proposition* 4.1 *also holds if* $\mathcal{IR}_{\mathcal{B}}$ *is full dimensional.*

Observation 4.3 *Let a f.c.b.* $\mathcal{B}$ *and an index* $i \in \mathcal{B}$ *be given such that* $\hbar^i_{\mathcal{B}}$ *forms a* $(k-1)$*-dimensional boundary of* $\mathcal{IR}_{\mathcal{B}}$*. It is clear from* (4.6) *that if* $j \in H^i_{\mathcal{B}}$, $\hbar^j_{\mathcal{B}}$ *also forms a* $(k-1)$*-dimensional boundary of* $\mathcal{IR}_{\mathcal{B}}$*.*

Together, Observations 4.1, 4.2, and 4.3 and Proposition 4.1 provide us with a strategy so that, given a f.c.b. $\mathcal{B}$ such that $\mathcal{IR}_{\mathcal{B}}$ is full dimensional, we can compute the set $\mathrm{F}_{\mathcal{B}}$. We present this strategy in Algorithm 2.

Note that the key step in Algorithm 2 is found on lines 2 and 3. Here we use NLP_F to identify the indices i in $\mathcal{B}$ for which $\hbar^i_{\mathcal{B}}$ forms a $(k-1)$-dimensional boundary of $\mathcal{IR}_{\mathcal{B}}$, i.e., i belongs to $\mathrm{F}_{\mathcal{B}}$.

The remainder of this chapter is devoted to the development of the theory necessary for execution of the subroutine GetAdjacentRegionsAcross. For this, we use the result of Proposition 3.11, which guarantees that for any pair of adjacent full dimensional invariancy regions $\mathcal{IR}_{\mathcal{B}_i}$ and $\mathcal{IR}_{\mathcal{B}_j}$ with $i < j$, there always exists a sequence of bases $\left\{\mathcal{B}_n\right\}_{n=i+1}^{j-1}$ such that $\mathcal{B}_\gamma$ and $\mathcal{B}_{\gamma+1}$ are adjacent

---

Algorithm 2 BuildF($\mathcal{B}$)—Build $\mathrm{F}_{\mathcal{B}}$

Input: A f.c.b. $\mathcal{B}$ such that $dim(\mathcal{IR}_{\mathcal{B}}) = k$.
Output: The set $\mathrm{F}_{\mathcal{B}}$.

1: Let $\mathrm{F}_{\mathcal{B}} = \emptyset$.
2: **for** $i \in \left(\mathcal{B} \setminus (\mathrm{Z}_{\mathcal{B}} \cup \mathrm{E}_{\mathcal{B}} \cup \mathrm{F}_{\mathcal{B}})\right)$ **do** solve $NLP_F(\mathcal{B}, i)$ to find an optimal solution $(\lambda^*, \phi^*, \upsilon^*)$.
3: **if** $\lambda^* > 0$ **then** add $\left(i \cup \mathrm{H}^i_{\mathcal{B}}\right)$ to $\mathrm{F}_{\mathcal{B}}$.
4: Return $\mathrm{F}_{\mathcal{B}}$.

---

for all $\gamma \in \{i, \ldots, j-1\}$ and $dim\left(\mathcal{IR}_{\mathcal{B}_\gamma}\right) \geq k-1$ for all $\gamma \in \{i+1, \ldots, j-1\}$. Recognize that there is a distinct difference between *adjacency of bases* and *adjacency of invariancy regions* and that one does not necessarily imply the other. Hence, the rest of this chapter is divided into two sections. In the first we discuss the computation of *adjacent bases*, and in the second we discuss the sequential use of adjacent basis calculations in order to find adjacent invariancy regions.

4.1 Computing Adjacent Bases

Here we establish a method so that given a complementary basis $\mathcal{B}$, we can determine a complementary basis $\mathcal{B}'$ adjacent to $\mathcal{B}$. Observe the following proposition that we obtain by extending the result of Lemma 3.8 of Columbano et al. [11].

Proposition 4.2 *If $M(\phi) \in \mathbb{R}^{h\times h}$ is column sufficient for each $\phi \in \Phi$ and two bases $\mathcal{B}_1$ and $\mathcal{B}_2$ are adjacent, then $\left|\mathcal{B}_1 \cap \mathcal{B}_2\right| \geq h-2$.*

Proposition 4.2 is powerful as it implies that given any basis $\mathcal{B}$, any complementary basis that is adjacent to $\mathcal{B}$ can be obtained by replacing either one or two elements of $\mathcal{B}$ with their complements.

Definition 4.1 Replacing a single element of a basis with its complement is called a *diagonal pivot*, and replacing two elements of a basis with their complements is called an *exchange pivot*.

These terms originate from solution procedures for nonparametric LCP that use a tableau representation of the LCP and employ principal pivoting to compute feasible solutions. For a given basis $\mathcal{B}$, the corresponding tableau is the augmented matrix

$$T_{\mathcal{B}}(\phi, \upsilon) := \left[\, G(\phi)^{-1}_{\bullet\mathcal{B}} G(\phi) \,\middle|\, G(\phi)^{-1}_{\bullet\mathcal{B}} q(\phi, \upsilon) \,\right]. \tag{4.8}$$

Note that the RHS of $T_{\mathcal{B}}(\phi, \upsilon)$ is $\nu_{\mathcal{B}}(\theta)$. Also, recognize that the elements of the tableau associated with a given basis $\mathcal{B}$ are rational functions of ϕ and υ. However, only the elements of the RHS of the tableau depend on υ. Thus, since the majority of the theory we present in this work for which we utilize tableaux deals with only the left hand side (LHS) elements, when appropriate, we drop the dependency of $T_{\mathcal{B}}$ on υ and use the notation $T_{\mathcal{B}}(\phi)$.

Given a basis $\mathcal{B}$, not every diagonal and exchange pivot will result in a new feasible basis. Thus, we need to determine conditions under which pivots will yield new adjacent bases. Such conditions can be developed using the tableau $T_{\mathcal{B}}(\phi, \upsilon)$. We first consider diagonal pivots. Since principal pivoting has been studied extensively in the context of LCP, the following result is a direct extension of a well-known fact in the literature.

Observation 4.4 *Given a complementary basis $\mathcal{B}$ and an $i \in \mathcal{B}$, the set $\big(\mathcal{B} \setminus \{i\}\big) \cup \{\bar{\imath}\}$ is a basis if and only if there exists $\phi \in \Phi$ such that $(T_{\mathcal{B}}(\phi))_{i,\bar{\imath}} \neq 0$.*

Observe the following proposition and its corollary, which are slightly adjusted from [11].

Proposition 4.3 *Given a complementary basis $\mathcal{B}$, suppose that for some index $i \in \mathcal{B}$ the set $\mathcal{B}' = \big(\mathcal{B} \setminus \{i\}\big) \cup \{\bar{\imath}\}$ is a basis. Then $\mathcal{B}$ and $\mathcal{B}'$ are adjacent.*

Proof For all $\phi \in \Phi$, we have $C_{\mathcal{B}}(\phi) \cap C_{\mathcal{B}'}(\phi) = cone\Big(G(\phi)_{\bullet,(\mathcal{B}\setminus\{i\})}\Big)$ and $dim\Big(cone\Big(G(\phi)_{\bullet,(\mathcal{B}\setminus\{i\})}\Big)\Big) = h - 1$. Therefore, by Definition 2.14, the bases $\mathcal{B}$ and $\mathcal{B}'$ are adjacent. □

Corollary 4.1 *Given a complementary basis $\mathcal{B}$, suppose that for some index $i \in \mathcal{B}$ the set $\mathcal{B}' = \big(\mathcal{B} \setminus \{i\}\big) \cup \{\bar{\imath}\}$ is a basis. If $M(\phi)$ is column sufficient for all $\phi \in \Phi$, then for each $\phi \in \Phi$, $C_{\mathcal{B}'}(\phi)$ is the unique parametric complementary cone adjacent to $C_{\mathcal{B}}(\phi)$ along the facet cone $\Big(G(\phi)_{\bullet,(\mathcal{B}\setminus\{i\})}\Big)$.*

Proof This is clear since for column sufficient matrices, the relative interiors of all complementary cones are disjoint. □

By combining the results of Observation 4.4 and Proposition 4.3, we now have a set of conditions so that, given a complementary basis $\mathcal{B}$, an adjacent complementary basis $\mathcal{B}'$ can be computed using a single diagonal pivot. We now introduce several lemmas and a new proposition that provides conditions so that, given a complementary basis $\mathcal{B}$, an adjacent complementary basis $\mathcal{B}'$ can be computed using a single exchange pivot. Note that aspects of one of the proofs in [11] are used in the proof that follows our proposition. Additionally, after our proposition, we include a corollary, the result of which is useful later in this section.

Lemma 4.1 (Proposition 6 of den Hertog et al. [27]) *Any principal pivotal transform of a sufficient matrix is sufficient.*

Lemma 4.2 *For any complementary basis $\mathcal{B}$, the matrix $(T_{\mathcal{B}}(\phi))_{\bullet,\bar{\mathcal{B}}}$ is sufficient.*

Proof The result follows from Lemma 4.1 and the facts that $M(\phi)$ is sufficient for all $\phi \in \Phi$ and $(T_{\mathcal{B}}(\phi))_{\bullet,\bar{\mathcal{B}}}$ is a principal pivotal transformation of $M(\phi)$. □

Lemma 4.3 *Suppose that for a given complementary basis $\mathcal{B}$ and some $\phi \in \Phi$, we have that $(T_{\mathcal{B}}(\phi))_{i,\bar{\imath}} = 0$. Then $(T_{\mathcal{B}}(\phi))_{j,\bar{\imath}} \neq 0$ if and only if $(T_{\mathcal{B}}(\phi))_{i,\bar{\jmath}} \neq 0$, and in this case $(T_{\mathcal{B}}(\phi))_{j,\bar{\imath}}(T_{\mathcal{B}}(\phi))_{i,\bar{\jmath}} < 0$.*

Proof The result follows directly from Lemma 4.2 and Proposition 5 of den Hertog et al. [27]. □

Proposition 4.4 *Let a complementary basis $\mathcal{B}$ and distinct $i, j \in \mathcal{B}$ be given. Given $\phi \in \text{Proj}_U \mathcal{IR}_{\mathcal{B}}$, the set $\mathcal{B}' = \big(\mathcal{B} \setminus \{i, j\}\big) \cup \{\bar{\imath}, \bar{\jmath}\}$ is a complementary basis such that $C_{\mathcal{B}}(\phi)$ is adjacent to $C_{\mathcal{B}'}(\phi)$ along the facet cone $\big(G(\phi)_{\bullet(\mathcal{B}\setminus\{i\})}\big)$ if and only if $(T_{\mathcal{B}}(\phi))_{i,\bar{\imath}} = 0$ and $(T_{\mathcal{B}}(\phi))_{j,\bar{\imath}} > 0$.*

Proof ($\Leftarrow$): We begin by showing that $\mathcal{B}'$ is a basis and later show that $\mathcal{C}_{\mathcal{B}}(\phi)$ is adjacent to $\mathcal{C}_{\mathcal{B}'}(\phi)$ along the facet $cone\left(G(\phi)_{\bullet(\mathcal{B}\setminus\{i\})}\right)$. Since the LHS of $T_{\mathcal{B}}(\phi)$ is given by $G(\phi)_{\bullet\mathcal{B}}^{-1}G(\phi)$, we know $(T_{\mathcal{B}}(\phi))_{i,\gamma} = 0$ for all $\gamma \in \mathcal{B} \setminus \{i\}$ and $(T_{\mathcal{B}}(\phi))_{j,\xi} = 0$ for all $\xi \in \mathcal{B} \setminus \{j\}$. Because we also know $(T_{\mathcal{B}}(\phi))_{j,\bar{\imath}} > 0$, this shows that $G(\phi)_{\bullet\bar{\imath}}$ is not a linear combination of the columns of $G(\phi)_{\bullet(\mathcal{B}\setminus\{i,j\})}$. We note that from Lemma 4.3 we also have that $(T_{\mathcal{B}}(\phi))_{i,\bar{\jmath}} \neq 0$. This, together with the fact that $(T_{\mathcal{B}}(\phi))_{i,\bar{\imath}} = 0$, shows that $G(\phi)_{\bullet\bar{\jmath}}$ is not a linear combination of the columns of $G(\phi)_{\bullet((\mathcal{B}\setminus\{i,j\})\cup\{\bar{\imath}\})}$. We can therefore conclude that the columns of $G(\phi)_{\bullet\mathcal{B}'}$ are linearly independent, i.e., $\mathcal{B}'$ is a basis.

Next we argue that $\mathcal{C}_{\mathcal{B}}(\phi)$ is adjacent to $\mathcal{C}_{\mathcal{B}'}(\phi)$ along the facet $cone\left(G(\phi)_{\bullet(\mathcal{B}\setminus\{i\})}\right)$. For any set $J \subset \mathcal{E}$ and $\upsilon \in \Upsilon$, we have that $q(\phi, \upsilon)$ is in the relative interior of $cone\,(G(\phi)_{\bullet J})$ if and only if $q(\phi, \upsilon)$ is a strictly positive combination of the columns of $G(\phi)_{\bullet J}$, i.e., for each $\gamma \in J$, there exists $\beta_\gamma > 0$ such that $q(\phi, \upsilon) = \sum\limits_{\gamma\in J}\beta_\gamma G(\phi)_{\bullet\gamma}$. Consider

$$\mathcal{W}(\beta) := \beta_j G(\phi)_{\bullet j} + \sum_{\gamma\in(\mathcal{B}\setminus\{i,j\})} \beta_\gamma G(\phi)_{\bullet\gamma}. \tag{4.9}$$

Since we have $(T_{\mathcal{B}}(\phi))_{\bullet\bar{\imath}} = G(\phi)_{\bullet\mathcal{B}}^{-1}G(\phi)_{\bullet\bar{\imath}}$, we know

$$G(\phi)_{\bullet\bar{\imath}} = G(\phi)_{\bullet j}(T_{\mathcal{B}}(\phi))_{j,\bar{\imath}} + \sum_{\gamma\in(\mathcal{B}\setminus\{i,j\})} G(\phi)_{\bullet\gamma}(T_{\mathcal{B}}(\phi))_{\gamma\bar{\imath}} \tag{4.10}$$

and consequently $G(\phi)_{\bullet j} = \dfrac{1}{(T_{\mathcal{B}}(\phi))_{j,\bar{\imath}}}G(\phi)_{\bullet\bar{\imath}} - \sum\limits_{\gamma\in(\mathcal{B}\setminus\{i,j\})} G(\phi)_{\bullet\gamma}\dfrac{(T_{\mathcal{B}}(\phi))_{\gamma\bar{\imath}}}{(T_{\mathcal{B}}(\phi))_{j,\bar{\imath}}}$ since $(T_{\mathcal{B}}(\phi))_{j,\bar{\imath}} > 0$. Substituting this result into (4.9) yields

$$\mathcal{W}(\beta) = \frac{\beta_j}{(T_{\mathcal{B}})_{j,\bar{\imath}}}G(\phi)_{\bullet\bar{\imath}} - \sum_{\gamma\in(\mathcal{B}\setminus\{i,j\})}\left(\beta_\gamma - \beta_j\frac{(T_{\mathcal{B}}(\phi))_{\gamma\bar{\imath}}}{(T_{\mathcal{B}}(\phi))_{j,\bar{\imath}}}\right)G(\phi)_{\bullet\gamma}. \tag{4.11}$$

Now, select $\tilde{\beta}$ with the following properties: (i) $\tilde{\beta}_\gamma > 0$ for all $\gamma \in \left(\mathcal{B} \setminus \{i\}\right)$, and (ii) $\tilde{\beta}_\gamma > \tilde{\beta}_j\dfrac{(T_{\mathcal{B}}(\phi))_{\gamma\bar{\imath}}}{(T_{\mathcal{B}}(\phi))_{j,\bar{\imath}}}$ for all $\gamma \in \left(\mathcal{B} \setminus \{i, j\}\right)$. Then by (4.9) and (4.11), we know $\mathcal{W}(\tilde{\beta})$ is in the relative interiors of both $cone\left(G(\phi)_{\bullet(\mathcal{B}\setminus\{i\})}\right)$ and $cone\left(G(\phi)_{\bullet(\mathcal{B}'\setminus\{\bar{\jmath}\})}\right)$. Hence, $dim(cone(G(\phi)_{\bullet(\mathcal{B}\setminus\{i\})}) \cap cone(G(\phi)_{\bullet(\mathcal{B}'\setminus\{\bar{\jmath}\})})) = h - 1$, and $\mathcal{B}$ and $\mathcal{B}'$ are adjacent.

($\Rightarrow$): For the proof of this direction, we employ contradiction. Consider the following cases:

Case 1: $(T_{\mathcal{B}}(\phi))_{i,\bar{\iota}} \neq 0$

We know from Corollary 4.1 that $\hat{\mathcal{B}} = (\mathcal{B} \setminus \{i\}) \cup \{\bar{\iota}\}$ is the unique basis for which $\mathcal{C}_{\hat{\mathcal{B}}}(\phi)$ is adjacent to $\mathcal{C}_{\mathcal{B}}(\phi)$ along the facet $cone\left(G(\phi)_{\bullet(\mathcal{B}\setminus\{i\})}\right)$. This is a contradiction.

Case 2: $(T_{\mathcal{B}}(\phi))_{j,\bar{\iota}} \leq 0$

Since $\mathcal{B}'$ is a basis, we know that (4.11) is the only way to represent $\mathcal{W}(\beta)$ as a linear combination of the columns of $G(\phi)_{\bullet\mathcal{B}'}$. As a result, there cannot exist a β such that $\mathcal{W}(\beta)$ is in both $\mathcal{C}_{\mathcal{B}'}(\phi)$ and the relative interior of $cone\left(G(\phi)_{\bullet(\mathcal{B}\setminus\{i\})}\right)$. Therefore, $\mathcal{C}_{\mathcal{B}}(\phi)$ is not adjacent to $\mathcal{C}_{\mathcal{B}'}(\phi)$ along $cone\left(G(\phi)_{\bullet(\mathcal{B}\setminus\{i\})}\right)$, which is a contradiction.

The contradictions found above show that we must have $(T_{\mathcal{B}}(\phi))_{i,\bar{\iota}} = 0$ and $(T_{\mathcal{B}}(\phi))_{j,\bar{\iota}} > 0$. □

Corollary 4.2 *Let complementary bases $\mathcal{B}$ and $\mathcal{B}'$ be given such that $\mathcal{B}' = \left(\mathcal{B} \setminus \{i, j\}\right) \cup \{\bar{\iota}, \bar{j}\}$ for distinct $i, j \in \mathcal{B}$. Then for any $\phi \in \Phi$ such that $(T_{\mathcal{B}}(\phi))_{i,\bar{\iota}} = 0$ and any $\upsilon \in \Upsilon$, the point $q(\phi, \upsilon)$ can only lie in both $\mathcal{C}_{\mathcal{B}'}(\phi)$ and the relative interior of $cone\left(G(\phi)_{\bullet(\mathcal{B}\setminus\{i\})}\right)$ if $(T_{\mathcal{B}}(\phi))_{j,\bar{\iota}} > 0$.*

Proof The result is a direct consequence of the contradiction found in the second case considered during the proof of the reverse direction of Proposition 4.4. □

Observe the following steps that combine the results of Propositions 4.2, 4.3, and 4.4, Observation 4.4, and Corollary 4.1 to form a strategy for finding all complementary bases that are adjacent to a given basis $\mathcal{B}$:

1. Calculate the tableau $T_{\mathcal{B}}(\phi)$ associated with basis $\mathcal{B}$.
2. For any $i \in \mathcal{B}$ for which there exists $\phi \in \Phi$ such that $(T_{\mathcal{B}}(\phi))_{i,\bar{\iota}} \neq 0$, the set $\left(\mathcal{B} \setminus \{i\}\right) \cup \{\bar{\iota}\}$ is a complementary basis adjacent to $\mathcal{B}$.
3. For any distinct $i, j \in \mathcal{B}$ for which there exists $\phi \in \mathrm{Proj}_U \mathcal{IR}_{\mathcal{B}}$ such that $(T_{\mathcal{B}}(\phi))_{i,\bar{\iota}} = 0$ and $(T_{\mathcal{B}}(\phi))_{j,\bar{\iota}} > 0$, the set $\left(\mathcal{B} \setminus \{i, j\}\right) \cup \{\bar{\iota}, \bar{j}\}$ is a complementary basis adjacent to $\mathcal{B}$.

We are now able to utilize this strategy in order to compute sequences of adjacent bases that link adjacent invariancy regions, as guaranteed by Proposition 3.11.

4.2 Computing Adjacent Invariancy Regions

In this section we present a technique that exploits the properties of adjacent bases discussed above in order to compute adjacent invariancy regions. However, this technique has many pieces and depends heavily on many foundational concepts. Hence, we further break down Section 4.2 as follows. In Section 4.2.1 we present theory and methodology for constructing the sets $\mathrm{E}_{\mathcal{B}}$ and $\mathrm{H}^i_{\mathcal{B}}$, given a f.c.b. $\mathcal{B}$. After this, we take a brief detour from considering adjacent invariancy region computation and use Section 4.2.2 to show how the theory developed so far applies to Examples 2.1 and 2.2. Section 4.2.3 contains a discussion on computing adjacent invariancy regions beginning with a full dimensional region. In Section 4.2.4 we no longer work under the assumption that a given invariancy region is full dimensional and thus present theory and methodology for determining the dimension of a given invariancy region. Finally, we examine the computation of adjacent invariancy regions, given a $(k-1)$-dimensional region, in Section 4.2.5.

4.2.1 Construction of Special Sets

Given a f.c.b. $\mathcal{B}$, we now consider the sets $\mathrm{Z}_{\mathcal{B}}$, $\mathrm{E}_{\mathcal{B}}$, and $\mathrm{H}^i_{\mathcal{B}}$. Recall that $\mathrm{Z}_{\mathcal{B}}$ is constructed easily by observing $T_{\mathcal{B}}(\phi, \upsilon)$. Thus, we now present the theory necessary for constructing the sets $\mathrm{E}_{\mathcal{B}}$ and $\mathrm{H}^i_{\mathcal{B}}$. The method for doing so is given by the routine BUILDEANDH, outlined in Algorithm 3. Note that when passing a f.c.b. $\mathcal{B}$ to the routine BUILDEANDH, we assume that the set $\mathrm{Z}_{\mathcal{B}}$ has already been constructed and is passed into the routine along with $\mathcal{B}$.

Proposition 4.5 *Given a f.c.b. $\mathcal{B}$ for which $dim(\mathcal{IR}_{\mathcal{B}}) \geq k-1$ and distinct indices $i, j \in \mathcal{B}$, $j \in H^i_{\mathcal{B}}$ if and only if the following nonlinear program has an optimal value of zero.*

$$
\begin{aligned}
NLP_H(\mathcal{B}, i, j) := \max_{\lambda,\phi,\upsilon} \quad & \lambda \\
\text{s.t.} \quad & g_{\mathcal{B}}\,(Adj(G(\phi)_{\boldsymbol{.}\mathcal{B}}))_{\ell\boldsymbol{.}}\, q(\phi, \upsilon) \geq 0 \quad \forall\, \ell \in \big(\mathcal{B} \setminus (Z_{\mathcal{B}} \cup \{i, j\})\big) \\
& (Adj(G(\phi)_{\boldsymbol{.}\mathcal{B}}))_{i\boldsymbol{.}}\, q(\phi, \upsilon) = 0 \\
& g_{\mathcal{B}}\,(Adj(G(\phi)_{\boldsymbol{.}\mathcal{B}}))_{j\boldsymbol{.}}\, q(\phi, \upsilon) \geq \lambda \\
& \phi \in \Phi, \;\; \upsilon \in \Upsilon
\end{aligned}
\tag{4.12}
$$

Proof ($\Rightarrow$): Let $(\lambda^*, \phi^*, \upsilon^*)$ be an optimal solution to $NLP_H(\mathcal{B}, i, j)$. Notice that since $j \in \mathrm{H}^i_{\mathcal{B}}$, every $\theta \in (\hbar^i_{\mathcal{B}} \cap \mathcal{IR}_{\mathcal{B}})$ is also in $(\hbar^j_{\mathcal{B}} \cap \mathcal{IR}_{\mathcal{B}})$. This means that for every $(\lambda, \phi, \upsilon)$ feasible to $NLP_H(\mathcal{B}, i, j)$, $(Adj(G(\phi)_{\boldsymbol{.}\mathcal{B}}))_{j\boldsymbol{.}}\, q(\phi, \upsilon) = 0$, which shows that $\lambda^* \leq 0$. Additionally, since $\dim(\mathcal{IR}_{\mathcal{B}}) \geq k-1$, there must exist $\theta' = (\phi', \upsilon')$ in $\mathcal{IR}_{\mathcal{B}}$, i.e., all defining inequalities of $\mathcal{IR}_{\mathcal{B}}$ are satisfied at θ'. Thus, $(0, \phi', \upsilon')$ is feasible to $NLP_H(\mathcal{B}, i, j)$. This shows that $\lambda^* \geq 0$. Thus, we must have $\lambda^* = 0$.
($\Leftarrow$): Suppose $(\lambda^*, \phi^*, \upsilon^*)$ is an optimal solution to $NLP_H(\mathcal{B}, i, j)$ and $\lambda^* = 0$. Recognize that if there existed $\theta' = (\phi', \upsilon')$ in $\mathcal{IR}_{\mathcal{B}}$ such that $\theta' \in \hbar^i_{\mathcal{B}}$ but $\theta' \notin \hbar^j_{\mathcal{B}}$, then there would also exist $\lambda' > 0$ such that $g_{\mathcal{B}}\,(Adj(G(\phi)_{\boldsymbol{.}\mathcal{B}}))_{j\boldsymbol{.}}\, q(\phi, \upsilon) = \lambda'$. Furthermore, $(\lambda', \phi', \upsilon')$ would be feasible for $NLP_H(\mathcal{B}, i, j)$. This contradicts the fact that $\lambda^* = 0$, though, and so we must have that for all $\theta \in (\hbar^i_{\mathcal{B}} \cap \mathcal{IR}_{\mathcal{B}})$, $\theta \in \hbar^j_{\mathcal{B}}$. Hence, $\left(\hbar^i_{\mathcal{B}} \cap \mathcal{IR}_{\mathcal{B}}\right) \subseteq \left(\hbar^j_{\mathcal{B}} \cap \mathcal{IR}_{\mathcal{B}}\right)$ and therefore $j \in \mathrm{H}^i_{\mathcal{B}}$. □

Proposition 4.6 *Given a f.c.b. $\mathcal{B}$ for which $dim(\mathcal{IR}_{\mathcal{B}}) \geq k-1$ and an index $i \in \mathcal{B}$, we have $i \in E_{\mathcal{B}}$ if and only if there exists $j \in \mathcal{B}$ such that $NLP_H(\mathcal{B}, i, j)$ has a strictly negative optimal value or is infeasible.*

Proof ($\Rightarrow$): Recall from (4.5) that since $i \in \mathrm{E}_{\mathcal{B}}$, we have $\hbar^i_{\mathcal{B}} \cap \mathcal{IR}_{\mathcal{B}} = \emptyset$. This shows that there does not exist a $j \in \mathcal{B}$ with a feasible solution $(\lambda^j, \phi^j, \upsilon^j)$ to $NLP_H(\mathcal{B}, i, j)$ for which $\lambda^j \geq 0$. Now, for each $j \in \mathcal{B}$, let $\overline{\mathcal{IR}}^j_{\mathcal{B}}$ represent the semi-algebraic superset of $\mathcal{IR}_{\mathcal{B}}$ that results from eliminating the defining constraint of $\mathcal{IR}_{\mathcal{B}}$ associated with index j from $\mathcal{IR}_{\mathcal{B}}$. For an arbitrary $j \in \mathcal{B}$, suppose that $\hbar^i_{\mathcal{B}} \cap \overline{\mathcal{IR}}^j_{\mathcal{B}} = \emptyset$. Clearly, in this case $NLP_H(\mathcal{B}, i, j)$ is infeasible. On the other hand, if $\hbar^i_{\mathcal{B}} \cap \overline{\mathcal{IR}}^j_{\mathcal{B}} \neq \emptyset$, let $\tilde{\theta}^j = (\tilde{\phi}^j, \tilde{\upsilon}^j)$ be a point in $\hbar^i_{\mathcal{B}} \cap \overline{\mathcal{IR}}^j_{\mathcal{B}}$. Then we have: (i) $g_{\mathcal{B}}\left(Adj(G(\tilde{\phi}^j)_{.\mathcal{B}})\right)_{\ell.} q(\tilde{\phi}^j, \tilde{\upsilon}^j) \geq 0$ for all $\ell \in \left(\mathcal{B} \setminus \{i, j\}\right)$, (ii) $\left(Adj(G(\tilde{\phi}^j)_{.\mathcal{B}})\right)_{i.} q(\tilde{\phi}^j, \tilde{\upsilon}^j) = 0$, and (iii) there exists $\tilde{\lambda}^j < 0$ such that $g_{\mathcal{B}}\left(Adj(G(\tilde{\phi}^j)_{.\mathcal{B}})\right)_{j.} q(\tilde{\phi}^j, \tilde{\upsilon}^j) \geq \tilde{\lambda}^j$. Thus, $NLP_H(\mathcal{B}, i, j)$ is feasible and has a strictly negative optimal value. We have now shown that for each $j \in \mathcal{B}$: (i) if $\hbar^i_{\mathcal{B}} \cap \overline{\mathcal{IR}}^j_{\mathcal{B}} = \emptyset$, then $NLP_H(\mathcal{B}, i, j)$ is infeasible, and (ii) if $\hbar^i_{\mathcal{B}} \cap \overline{\mathcal{IR}}^j_{\mathcal{B}} \neq \emptyset$, then $NLP_H(\mathcal{B}, i, j)$ is feasible but has a strictly negative optimal value.

($\Leftarrow$): Recognize that if there existed $\theta' = (\phi', \upsilon')$ in $\hbar^i_{\mathcal{B}} \cap \mathcal{IR}_{\mathcal{B}}$, then there would also exist $\lambda' \geq 0$ such that $g_{\mathcal{B}}(Adj(G(\phi)_{.\mathcal{B}}))_{\ell.} q(\phi, \upsilon) = \lambda'$ for all $\ell \in \mathcal{B}$. Furthermore, $(\lambda', \phi', \upsilon')$ would be feasible to $NLP_H(\mathcal{B}, i, \ell)$ for all $\ell \in \mathcal{B}$. However, this contradicts the fact that there exists $j \in \mathcal{B}$ for which the optimal value of $NLP_H(\mathcal{B}, i, j)$ is strictly negative. So, we must have that $(\hbar^i_{\mathcal{B}} \cap \mathcal{IR}_{\mathcal{B}}) = \emptyset$, which shows that $i \in \mathrm{E}_{\mathcal{B}}$. □

The results of Propositions 4.5 and 4.6 provide a strategy so that, given a f.c.b. $\mathcal{B}$ together with the set $\mathrm{Z}_{\mathcal{B}}$, we can build the sets $\mathrm{E}_{\mathcal{B}}$ and $\mathrm{H}^i_{\mathcal{B}}$ for each $i \in \mathcal{B}$. We present this strategy in Algorithm 3.

The key steps of Algorithm 3 are contained on lines 5–7. Lines 5 and 6 utilize NLP_H and the result of Proposition 4.5 to identify the indices in $\mathcal{B}$ that are present

Algorithm 3 BUILDEANDH($\mathcal{B}$)—Build $\mathrm{E}_{\mathcal{B}}$ and $\mathrm{H}^i_{\mathcal{B}}$ for each $i \in \mathcal{B}$

Input: A f.c.b. $\mathcal{B}$ such that $dim(\mathcal{IR}_{\mathcal{B}}) \geq k-1$.
Output: The sets $\mathrm{E}_{\mathcal{B}}$ and $\mathrm{H}^i_{\mathcal{B}}$ for each $i \in \mathcal{B}$.

1: Let $\mathrm{E}_{\mathcal{B}} = \emptyset$.
2: Let $\mathrm{H}^\ell_{\mathcal{B}} = \emptyset$ for each $\ell \in \mathcal{B}$.
3: **for** $i \in \left(\mathcal{B} \setminus (\mathrm{Z}_{\mathcal{B}} \cup \mathrm{E}_{\mathcal{B}})\right)$ **do**
4:     **for** $j \in \left(\mathcal{B} \setminus (\mathrm{Z}_{\mathcal{B}} \cup \mathrm{E}_{\mathcal{B}} \cup \{i\})\right)$ **do**
5:         **if** $j \notin \mathrm{H}^i_{\mathcal{B}}$ **then** solve $NLP_H(\mathcal{B}, i, j)$ to obtain an optimal solution $(\lambda^*, \phi^*, \upsilon^*)$.
6:             **if** $\lambda^* = 0$ **then** add $\left(j \cup \mathrm{H}^j_{\mathcal{B}}\right)$ to $\mathrm{H}^i_{\mathcal{B}}$.
7:             **else if** $\lambda^* < 0$ **then** add i to $\mathrm{E}_{\mathcal{B}}$ and exit the **for** loop beginning on Line 4.
8: Return $\mathrm{E}_{\mathcal{B}}$ and $\mathrm{H}^\ell_{\mathcal{B}}$ for each $\ell \in \mathcal{B}$.

in $H^i_{\mathcal{B}}$, for each $i \in \mathcal{B}$. Line 7 utilizes the result of Proposition 4.6 to identify the indices of $\mathcal{B}$ that are present in $E_{\mathcal{B}}$. Recognize that by using the convention that the optimal value of an infeasible maximization problem is $-\infty$, line 7 also covers the case in which NLP_H is infeasible.

We now break from our discussion on the computation of adjacent invariancy regions and show how the theory presented so far applies to Examples 2.1 and 2.2.

4.2.2 *Revisiting Examples*

Throughout the remainder of this section, we introduce several more propositions whose results allow us to develop the procedure GETADJACENTREGIONSACROSS. Similar to Propositions 4.1 and 4.5, many of these propositions introduce nonlinear programs (NLPs) that can be solved in order to determine, for example, if a given invariancy region is full dimensional. We note that many of these NLPs do not need to be solved to optimality, but rather a feasible solution must be found that has an associated objective function value that is strictly positive. As we develop the theory necessary for partitioning Θ, we also show directly how this theory can be applied to each of the instances we introduced earlier in Examples 2.1 and 2.2. We will do so, though, using the subvectors ϕ and υ rather than θ. Consider (2.1) and (2.2) and recall that ϕ represents the subvector of θ such that every element of ϕ is present in some element of $M(\theta)$ and υ represents the subvector of θ such that no element of υ is present in any element of $M(\theta)$. Hence, we have: (i) for Example 2.1 , $\phi = \begin{bmatrix} \theta_1 \\ \theta_2 \end{bmatrix}$ and $\upsilon = \emptyset$, and (ii) for Example 2.2, $\phi = \theta_1$ and $\upsilon = \theta_2$. For the sake of clarity, when discussing these examples we use variable names to describe the elements of each basis rather than their corresponding indices. The solution to Example 2.1 is given in Table 4.1 and depicted in Figure 4.1a, while the solution to Example 2.2 is given in Table 4.2 and depicted in Figure 4.1b. How each solution is determined is shown during the discussion that follows. We note that each nonlinear program solved throughout the course of this work is solved using the "fmincon" function in MATLAB.

Due to the size of the tableaux we utilize throughout this chapter and Chapter 5, we do not include them in their entirety in these chapters. Instead, we include tableaux for Examples 2.1 and 2.2 in Appendices A and B, respectively. We note that for Example 2.2, phase 1 of our procedure requires only one iteration and, thus, all tables in Appendix B are dedicated to phase 2. On the other hand, for Example 2.1, phase 1 requires several iterations and for this reason Tables A.1–A.9 are dedicated to phase 1 and Table A.10 is dedicated to phase 2. Note that although Tables A.6 and A.7 display tableaux for f.c.b.'s initially discovered during phase 1, both are reused in phase 2. Additionally, the sizes of several of the tableaux for Example 2.1 are so large that we cannot include every column. In these cases we include only the columns associated with nonbasic variables (note that the columns

Table 4.1 Solution for Example 2.1

$$\mathcal{B}_0^{2.1}: \quad \begin{bmatrix} w_1 = \dfrac{3\phi_1^3+18\phi_1^2\phi_2-49\phi_1^2-75\phi_1\phi_2^2+148\phi_1\phi_2+68\phi_1+96\phi_2^2-16\phi_2-76}{2\left(-3\phi_1^2+8\phi_1\phi_2+19\phi_1+41\phi_2^2-24\phi_2-22\right)} \\ z_2 = -\dfrac{(\phi_1+2)\left(9\phi_1^3-9\phi_1^2\phi_2-33\phi_1^2-87\phi_1\phi_2^2+21\phi_1\phi_2+22\phi_1-59\phi_2^3+13\phi_2^2+50\phi_2+5\right)}{4\left(-3\phi_1^2+8\phi_1\phi_2+19\phi_1+41\phi_2^2-24\phi_2-22\right)} \\ z_3 = \dfrac{(\phi_1+2)\left(-6\phi_1^2-\phi_1\phi_2+11\phi_1+15\phi_2^2-16\phi_2+1\right)}{2\left(-3\phi_1^2+8\phi_1\phi_2+19\phi_1+41\phi_2^2-24\phi_2-22\right)} \\ z_4 = \dfrac{(\phi_1+2)\left(3\phi_1^2+8\phi_1\phi_2-\phi_1+5\phi_2^2+5\phi_2-11\right)}{2\left(-3\phi_1^2+8\phi_1\phi_2+19\phi_1+41\phi_2^2-24\phi_2-22\right)} \\ z_5 = \dfrac{(\phi_1+2)\left(9\phi_1-13\phi_2+\phi_1\phi_2+21\phi_2^2-12\right)}{2\left(-3\phi_1^2+8\phi_1\phi_2+19\phi_1+41\phi_2^2-24\phi_2-22\right)} \end{bmatrix}$$

$$\mathcal{B}_1^{2.1}: \quad \begin{bmatrix} z_1 = \dfrac{3\phi_1^3+18\phi_1^2\phi_2-49\phi_1^2-75\phi_1\phi_2^2+148\phi_1\phi_2+68\phi_1+96\phi_2^2-16\phi_2-76}{8(32\phi_2-7\phi_1+29)} \\ z_2 = \dfrac{-81\phi_1^3+288\phi_1^2\phi_2-111\phi_1^2+47\phi_1\phi_2^2+856\phi_1\phi_2+422\phi_1+544\phi_2^2+392\phi_2-44}{16(32\phi_2-7\phi_1+29)} \\ z_3 = -\dfrac{61\phi_1-60\phi_1\phi_2+29\phi_1^2-12}{4(32\phi_2-7\phi_1+29)} \\ z_4 = \dfrac{75\phi_1+32\phi_2-5\phi_1\phi_2+13\phi_1^2+20}{4(32\phi_2-7\phi_1+29)} \\ z_5 = \dfrac{16\phi_1+96\phi_2+9\phi_1\phi_2+2\phi_1^2+84}{4(32\phi_2-7\phi_1+29)} \end{bmatrix}$$

$$\mathcal{B}_2^{2.1}: \quad \begin{bmatrix} w_1 = \dfrac{-11\phi_1^2+37\phi_1+48\phi_2-44}{2(7\phi_1+8\phi_2-13)} \\ z_2 = -\dfrac{(\phi_1+2)\left(-9\phi_1^2-23\phi_1\phi_2+15\phi_1-17\phi_2^2+23\phi_2+3\right)}{4(7\phi_1+8\phi_2-13)} \\ w_3 = -\dfrac{(\phi_1+2)\left(-6\phi_1^2-\phi_1\phi_2+11\phi_1+15\phi_2^2-16\phi_2+1\right)}{2(7\phi_1+8\phi_2-13)} \\ z_4 = \dfrac{(\phi_1+2)(3\phi_1+5\phi_2-6)}{2(7\phi_1+8\phi_2-13)} \\ z_5 = \dfrac{(\phi_1+2)(4\phi_1+3\phi_2-7)}{2(7\phi_1+8\phi_2-13)} \end{bmatrix}$$

for basic variables are identity vectors). We also point out that each of the tables in Appendices A and B is not necessarily obtained from the previous table. For example, bases $\mathcal{B}_i^{2.1}$, $\mathcal{B}_{iv}^{2.1}$, and $\mathcal{B}_{viii}^{2.1}$ can all be obtained from exchange pivots from $\mathcal{B}^{*2.1}$ and thus Tables A.2, A.5, and A.9 can all be obtained from Table A.1. As another example, $\mathcal{B}_2^{2.1}$ can be obtained from a diagonal pivot from basis $\mathcal{B}_0^{2.1}$ and so Table A.6 can be obtained from Table A.7. How each table is obtained from previous tables will be made clear as we consider Examples 2.1 and 2.2 in more detail in this chapter and in Chapter 5.

For Examples 2.1 and 2.2, we claim that the initial bases $\mathcal{B}_0^{2.1} = \{w_1, z_2, z_3, z_4, z_5\}$ and $\mathcal{B}_0^{2.2} = \{w_1, w_2, w_3, w_4\}$ are feasible. Respective tableaux for $\mathcal{B}_0^{2.1}$ and $\mathcal{B}_0^{2.2}$ are contained in Tables A.7 and B.1, which are found in

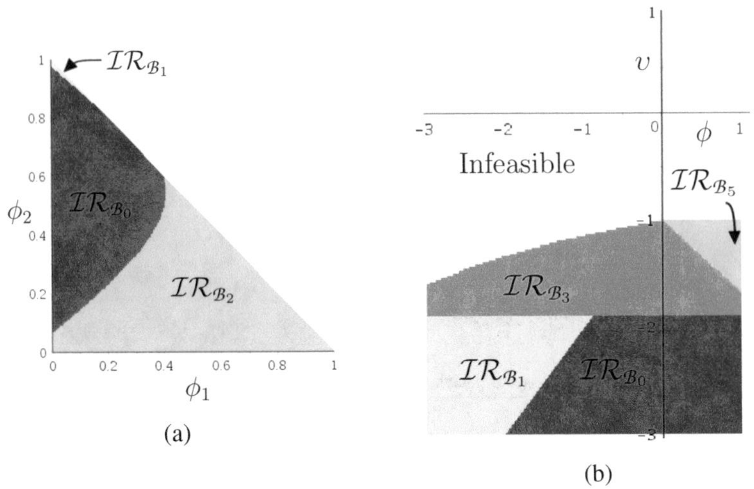

Fig. 4.1 Partitions of Θ for the two examples. (**a**) Example 2.1. (**b**) Example 2.2

Table 4.2 Solution for Example 2.2

$\mathcal{B}_0^{2.2}$:	$\begin{bmatrix} w_1 = -\upsilon - 1 \\ w_2 = \phi - \upsilon - 1 \\ w_3 = -18\,\upsilon - 34 \\ w_4 = -9\,\upsilon - 17 \end{bmatrix}$	$\mathcal{B}_1^{2.2}$:	$\begin{bmatrix} w_1 = -\frac{8\,\upsilon+\phi\,\upsilon+8}{\phi+7} \\ z_2 = -\frac{9\,\upsilon+17}{\phi+5} \\ w_3 = -\frac{(2\,\phi+13)\,(9\,\upsilon+17)}{\phi+5} \\ z_4 = \frac{\upsilon-\phi+1}{\phi+7} \end{bmatrix}$
$\mathcal{B}_3^{2.2}$:	$\begin{bmatrix} z_1 = \frac{(2\,\phi+13)\,(9\,\upsilon+17)}{\phi+8} \\ z_2 = \frac{9\,\upsilon+17}{\phi+8} \\ z_3 = -\frac{8\,\upsilon+\phi\,\upsilon+8}{2\,\phi+19} \\ z_4 = -\frac{\phi+\frac{3}{2}\upsilon+\frac{3}{2}}{\phi+\frac{19}{2}} \end{bmatrix}$	$\mathcal{B}_5^{2.2}$:	$\begin{bmatrix} z_1 = 18\,\upsilon + 34 \\ w_2 = \phi + \frac{3}{2}\upsilon + \frac{3}{2} \\ z_3 = -\frac{1}{2}\upsilon - \frac{1}{2} \\ w_4 = 9\,\upsilon + 17 \end{bmatrix}$

Appendices A and B, respectively. A discussion on obtaining initial bases is provided in Chapter 5.

We now show the use of the procedure BUILDEANDH (Algorithm 3) to construct $\mathrm{E}_{\mathcal{B}_0^{2.1}}$, $\mathrm{H}^i_{\mathcal{B}_0^{2.1}}$ for each $i \in \mathcal{B}_0^{2.1}$, $\mathrm{E}_{\mathcal{B}_0^{2.2}}$ and $\mathrm{H}^j_{\mathcal{B}_0^{2.2}}$ for each $j \in \mathcal{B}_0^{2.2}$. Afterward, we use BUILDF (Algorithm 2) to create $F_{\mathcal{B}_0^{2.1}}$ and $F_{\mathcal{B}_0^{2.2}}$. From (4.4) and Tables A.7 and B.1, recognize that $\mathrm{Z}_{\mathcal{B}_0^{2.1}} = \mathrm{Z}_{\mathcal{B}_0^{2.2}} = \emptyset$. Additionally, upon initial execution of BUILDEANDH, $\mathrm{E}_{\mathcal{B}_0^{2.1}} = \mathrm{E}_{\mathcal{B}_0^{2.2}} = \emptyset$. Hence, the first nontrivial step in executing BUILDEANDH($\mathcal{B}_0^{2.1}$) is to solve $NLP_H(\mathcal{B}_0^{2.1}, w_1, z_2)$:

$$\begin{aligned} \max_{\lambda,\phi} \quad & \lambda \\ \text{s.t.} \quad & -3\phi_1^3 - 18\phi_1^2\phi_2 + 49\phi_1^2 + 75\phi_1\phi_2^2 - 148\phi_1\phi_2 - 68\phi_1 - 96\phi_2^2 + 16\phi_2 + 76 = 0 \\ & (\phi_1+2)(9\phi_1^3 - 9\phi_1^2\phi_2 - 33\phi_1^2 - 87\phi_1\phi_2^2 + 21\phi_1\phi_2 + 22\phi_1 - 59\phi_2^3 + 13\phi_2^2 + 50\phi_2 + 5) \geq \lambda \\ & (\phi_1+2)(6\phi_1^2 + \phi_1\phi_2 - 11\phi_1 - 15\phi_2^2 + 16\phi_2 - 1) \geq 0 \\ & (\phi_1+2)(-3\phi_1^2 - 8\phi_1\phi_2 + \phi_1 - 5\phi_2^2 - 5\phi_2 + 11) \geq 0 \\ & (\phi_1+2)(-9\phi_1 + 13\phi_2 - \phi_1\phi_2 - 21\phi_2^2 + 12) \geq 0 \\ & \phi_1 + \phi_2 \leq 1 \\ & \phi_1, \phi_2 \geq 0 \end{aligned}$$

An approximate optimal solution to $NLP_H(\mathcal{B}_0^{2.1}, w_1, z_2)$ is $(\lambda^*, \phi^*, \upsilon^*) = (34.8687, 0.1478, 0.8522)$, which, based on the result of Proposition 4.5, shows that $z_2 \notin \mathrm{H}^{w_1}_{\mathcal{B}_0^{2.1}}$. For the sake of space, we do not explicitly show the additional ten NLPs solved during the BUILDH($\mathcal{B}_0^{2.1}$) procedure, but we do provide the following outline of the results:

w_1: z_2 – Optimal value of $NLP_H(\mathcal{B}_0^{2.1}, w_1, z_2) \approx 34.8687$. Do not add z_2 to $\mathrm{E}_{\mathcal{B}_0^{2.1}}$ or $\mathrm{H}^{w_1}_{\mathcal{B}_0^{2.1}}$.
z_3 – Optimal value of $NLP_H(\mathcal{B}_0^{2.1}, w_1, z_3) \approx 0.8015$. Do not add z_3 to $\mathrm{E}_{\mathcal{B}_0^{2.1}}$ or $\mathrm{H}^{w_1}_{\mathcal{B}_0^{2.1}}$.
z_4 – Optimal value of $NLP_H(\mathcal{B}_0^{2.1}, w_1, z_4) \approx 4.6874$. Do not add z_4 to $\mathrm{E}_{\mathcal{B}_0^{2.1}}$ or $\mathrm{H}^{w_1}_{\mathcal{B}_0^{2.1}}$.
z_5 – Optimal value of $NLP_H(\mathcal{B}_0^{2.1}, w_1, z_5) \approx 16.1191$. Do not add z_5 to $\mathrm{E}_{\mathcal{B}_0^{2.1}}$ or $\mathrm{H}^{w_1}_{\mathcal{B}_0^{2.1}}$.
z_2:w_1 – $NLP_H(\mathcal{B}_0^{2.1}, z_2, w_1)$ is infeasible. Add z_2 to $\mathrm{E}_{\mathcal{B}_0^{2.1}}$ and cease consideration of z_2.
z_3:w_1 – Optimal value of $NLP_H(\mathcal{B}_0^{2.1}, z_3, w_1) \approx 76.64$. Do not add w_1 to $\mathrm{E}_{\mathcal{B}_0^{2.1}}$ or $\mathrm{H}^{z_3}_{\mathcal{B}_0^{2.1}}$.
z_4 – Optimal value of $NLP_H(\mathcal{B}_0^{2.1}, z_3, z_4) \approx 21.348$. Do not add z_4 to $\mathrm{E}_{\mathcal{B}_0^{2.1}}$ or $\mathrm{H}^{z_3}_{\mathcal{B}_0^{2.1}}$.
z_5 – Optimal value of $NLP_H(\mathcal{B}_0^{2.1}, z_3, z_5) \approx 30.1402$. Do not add z_5 to $\mathrm{E}_{\mathcal{B}}$ or $\mathrm{H}^{z_3}_{\mathcal{B}_0^{2.1}}$.
z_4:w_1 – $NLP_H(\mathcal{B}_0^{2.1}, z_4, w_1)$ is infeasible. Add z_4 to $\mathrm{E}_{\mathcal{B}_0^{2.1}}$ and cease consideration of z_4.
z_5:w_1 – $NLP_H(\mathcal{B}_0^{2.1}, z_5, w_1)$ is infeasible. Add z_5 to $\mathrm{E}_{\mathcal{B}_0^{2.1}}$ and cease consideration of z_5.

After running BUILDEANDH($\mathcal{B}_0^{2.1}$), we find that $\mathrm{E}_{\mathcal{B}_0^{2.1}} = \{z_2, z_4, z_5\}$ and $\mathrm{H}^i_{\mathcal{B}_0^{2.1}} = \emptyset$ for all $i \in \mathcal{B}$. We do not explicitly show the steps of running BUILDEANDH($\mathcal{B}_0^{2.2}$), but the results reveal that $\mathrm{E}_{\mathcal{B}_0^{2.2}} = \{w_1\}$, $\mathrm{H}^{w_1}_{\mathcal{B}_0^{2.1}} = \mathrm{H}^{w_2}_{\mathcal{B}_0^{2.1}} = \emptyset$, $\mathrm{H}^{w_3}_{\mathcal{B}_0^{2.1}} = \{w_4\}$, and $\mathrm{H}^{w_4}_{\mathcal{B}_0^{2.1}} = \{w_3\}$. One can observe from Figure 4.2a and b that these sets have been constructed correctly.

We are now ready to use the procedure BUILDF to construct the sets $\mathrm{F}_{\mathcal{B}_0^{2.1}}$ and $\mathrm{F}_{\mathcal{B}_0^{2.2}}$. For both examples, we first consider the index w_1. Recall, however, that $w_1 \in \mathrm{E}_{\mathcal{B}_0^{2.2}}$, and thus, by Observation 4.1, we can immediately conclude that $w_1 \notin \mathrm{F}_{\mathcal{B}_0^{2.2}}$. Hence, we need to observe only $NLP_F(\mathcal{B}_0^{2.1}, w_1)$ (4.7):

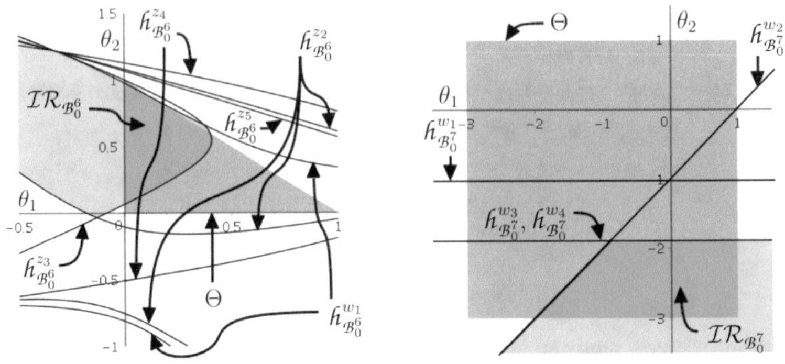

(a) (b)

Fig. 4.2 Hypersurfaces associated with initial bases for the two examples. (**a**) Hypersurfaces for $\mathcal{B}^{2.1}_0$. (**b**) Hypersurfaces for $\mathcal{B}^{2.2}_0$

$$
\begin{array}{ll}
\max\limits_{\lambda,\phi} & \lambda \\
\text{s.t.} & -3\phi_1^3 - 18\phi_1^2\phi_2 + 49\phi_1^2 + 75\phi_1\phi_2^2 - 148\phi_1\phi_2 - 68\phi_1 - 96\phi_2^2 + 16\phi_2 + 76 = 0 \\
& (\phi_1+2)(9\phi_1^3 - 9\phi_1^2\phi_2 - 33\phi_1^2 - 87\phi_1\phi_2^2 + 21\phi_1\phi_2 + 22\phi_1 - 59\phi_2^3 + 13\phi_2^2 + 50\phi_2 + 5) \geq \lambda \\
& (\phi_1+2)(6\phi_1^2 + \phi_1\phi_2 - 11\phi_1 - 15\phi_2^2 + 16\phi_2 - 1) \geq \lambda \\
& (\phi_1+2)(-3\phi_1^2 - 8\phi_1\phi_2 + \phi_1 - 5\phi_2^2 - 5\phi_2 + 11) \geq \lambda \\
& (\phi_1+2)(-9\phi_1 + 13\phi_2 - \phi_1\phi_2 - 21\phi_2^2 + 12) \geq \lambda \\
& \phi_1 + \phi_2 \leq 1 \\
& \phi_1, \phi_2 \geq 0
\end{array}
$$

An approximate optimal solution is $(\lambda^*, \phi_1^*, \phi_2^*) = (0.8015, 0.1388, 0.8599)$. Thus, by Proposition 4.1 and using (4.1), we see that $h^{w_1}_{\mathcal{B}^{2.1}_0} = \{\phi \in \Phi : -3\phi_1^3 - 18\phi_1^2\phi_2 + 49\phi_1^2 + 75\phi_1\phi_2^2 - 148\phi_1\phi_2 - 68\phi_1 - 96\phi_2^2 + 16\phi_2 + 76 = 0\}$ forms a $(k-1)$-dimensional boundary of $\mathcal{IR}_{\mathcal{B}^{2.1}_0}$. Completing the procedures BUILDF$\left(\mathcal{B}^{2.1}_0\right)$ and BUILDF$\left(\mathcal{B}^{2.2}_0\right)$ shows that $h^{z_3}_{\mathcal{B}^{2.1}_0} = \{\phi \in \Phi : (\phi_1+2)(6\phi_1^2 + \phi_1\phi_2 - 11\phi_1 - 15\phi_2^2 + 16\phi_2 - 1) = 0\}$ forms the only additional $(k-1)$-dimensional boundary of $\mathcal{IR}_{\mathcal{B}^{2.1}_0}$ and $h^{w_2}_{\mathcal{B}^{2.2}_0} = \{\theta \in \Theta : \phi - \upsilon - 1 = 0, \theta = (\phi, \upsilon)\}$ and $h^{w_3}_{\mathcal{B}^{2.2}_0} = h^{w_4}_{\mathcal{B}^{2.2}_0} = \{\theta \in \Theta : -9\upsilon - 17 = 0, \phi \in \Phi, \theta = (\phi, \upsilon)\}$ form the $(k-1)$-dimensional boundaries of $\mathcal{IR}_{\mathcal{B}^{2.2}_0}$. Hence, $\mathrm{F}_{\mathcal{B}^{2.1}_0} = \{w_1, z_3\}$ and $\mathrm{F}_{\mathcal{B}^{2.2}_0} = \{w_2, w_3, w_4\}$. To see that the correct conclusions have been made about which surfaces form $(k-1)$-dimensional boundaries of $\mathcal{IR}_{\mathcal{B}^{2.1}_0}$ and $\mathcal{IR}_{\mathcal{B}^{2.2}_0}$, observe Figure 4.2, which shows $\mathcal{IR}_{\mathcal{B}^{2.1}_0}$, $\mathcal{IR}_{\mathcal{B}^{2.2}_0}$, and the hypersurfaces associated with each basic variable for Examples 2.1 and 2.2.

We now suspend our consideration of Examples 2.1 and 2.2 in order to develop more of the theoretical results needed for implementation of the routine GETAD-

JACENTREGIONSACROSS. As we progress through the material that follows, we will frequently pause to demonstrate the application of newly developed concepts to Examples 2.1 and 2.2.

4.2.3 Determining Adjacent Invariancy Regions Given a Full Dimensional Region

We now continue our discussion and develop the tools necessary for determining bases having invariancy regions that are adjacent to a given full dimensional invariancy region across a known $(k-1)$-dimensional boundary. This discussion culminates with the presentation of Algorithm 4, which outlines the procedure GETADJACENTREGIONSACROSS. Consider the following definition, lemma, proposition, and corollary.

Definition 4.2 Given a complementary basis $\mathcal{B}$, the associated tableau $T_{\mathcal{B}}(\phi)$, and distinct indices $i \in \mathcal{B}$ and $j \in \mathcal{E}$, a *pivot on* $(T_{\mathcal{B}}(\phi))_{i,j}$ is the process of creating a new matrix T^* by performing elementary row operations on $T_{\mathcal{B}}(\phi)$ so that $T^*_{i,j} = 1$ and $T^*_{\gamma,j} = 0$ for all $\gamma \in (\mathcal{B} \setminus \{i\})$.

Note that for any basis $\mathcal{B}$ a pivot on $(T_{\mathcal{B}}(\phi))_{i,j}$ can only be made if $(T_{\mathcal{B}}(\phi))_{i,j} \not\equiv 0$.

Lemma 4.4 *Let distinct f.c.b.'s $\mathcal{B}$ and $\mathcal{B}'$ be given such that for some $i, j \in \mathcal{B}$, we have $\mathcal{B}' = (\mathcal{B} \setminus \{i, j\}) \cup \{\bar{\imath}, \bar{\jmath}\}$. Then for all $\phi \in \operatorname{Proj}_U \mathcal{IR}_{\mathcal{B}}$ such that $\mathcal{C}_{\mathcal{B}}(\phi)$ is adjacent to $\mathcal{C}_{\mathcal{B}'}(\phi)$ along the facet cone $\left(G(\phi)_{\bullet(\mathcal{B}\setminus\{i\})}\right)$, the RHS elements of $T_{\mathcal{B}'}(\phi)$ associated with $\bar{\imath}$ and $\bar{\jmath}$ are*

$$\frac{1}{(T_{\mathcal{B}}(\phi))_{i,\bar{\jmath}}(T_{\mathcal{B}}(\phi))_{j,\bar{\imath}}}\left((T_{\mathcal{B}}(\phi))_{i,\bar{\jmath}}\left(G(\phi)^{-1}_{\bullet\mathcal{B}}\right)_{j\bullet} - (T_{\mathcal{B}}(\phi))_{j,\bar{\jmath}}\left(G(\phi)^{-1}_{\bullet\mathcal{B}}\right)_{i\bullet}\right)q(\phi, \upsilon)$$

and

$$\frac{\left(G(\phi)^{-1}_{\bullet\mathcal{B}}\right)_{i\bullet}q(\phi, \upsilon)}{(T_{\mathcal{B}}(\phi))_{i,\bar{\jmath}}},$$

respectively.

Proof Let $\phi' \in \operatorname{Proj}_U \mathcal{IR}_{\mathcal{B}}$ be given such that $\mathcal{C}_{\mathcal{B}}(\phi')$ is adjacent to $\mathcal{C}_{\mathcal{B}'}(\phi')$ along the facet *cone* $\left(G(\phi')_{\bullet(\mathcal{B}\setminus\{i\})}\right)$. Then by Proposition 4.4, we have that $(T_{\mathcal{B}}(\phi'))_{i,\bar{\imath}} = 0$ and $(T_{\mathcal{B}}(\phi'))_{j,\bar{\imath}} > 0$. Furthermore, by Lemma 4.3 we also know that $(T_{\mathcal{B}}(\phi'))_{i,\bar{\jmath}} < 0$. Therefore, the tableau $T_{\mathcal{B}'}(\phi')$ can be obtained from the tableau $T_{\mathcal{B}}(\phi')$ in two steps: (i) create matrix T^* from $T_{\mathcal{B}}(\phi')$ by performing a pivot on $(T_{\mathcal{B}}(\phi'))_{j,\bar{\imath}}$ and (ii) obtain $T_{\mathcal{B}'}(\phi')$ from T^* by performing a pivot on $(T^*)_{i,\bar{\jmath}}$. Consider the following subset of $T_{\mathcal{B}}(\phi')$:

	i	j	$\bar{\imath}$	$\bar{\jmath}$	
i	1	0	0	$(T_{\mathcal{B}}(\phi'))_{i,\bar{\jmath}}$	$\left(G(\phi')^{-1}_{\bullet\mathcal{B}}\right)_{i\bullet} q(\phi', \upsilon)$
γ	0	0	$(T_{\mathcal{B}}(\phi'))_{\gamma,\bar{\imath}}$	$(T_{\mathcal{B}}(\phi'))_{\gamma,\bar{\jmath}}$	$\left(G(\phi')^{-1}_{\bullet\mathcal{B}}\right)_{\gamma\bullet} q(\phi', \upsilon)$
j	0	1	$(T_{\mathcal{B}}(\phi'))_{j,\bar{\imath}}$	$(T_{\mathcal{B}}(\phi'))_{j,\bar{\jmath}}$	$\left(G(\phi')^{-1}_{\bullet\mathcal{B}}\right)_{j\bullet} q(\phi', \upsilon)$

where γ represents any element of $\mathcal{B} \setminus \{i, j\}$. Pivoting on $(T_{\mathcal{B}}(\phi'))_{j,\bar{\imath}}$ yields the following corresponding subset of T^*:

	j	$\bar{\jmath}$	
i	0	$(T_{\mathcal{B}}(\phi'))_{i,\bar{\jmath}}$	$\left(G(\phi')^{-1}_{\bullet\mathcal{B}}\right)_{i\bullet} q(\phi', \upsilon)$
γ	$-\frac{(T_{\mathcal{B}}(\phi'))_{\gamma,\bar{\imath}}}{(T_{\mathcal{B}}(\phi'))_{j,\bar{\imath}}}$	$(T_{\mathcal{B}}(\phi'))_{\gamma,\bar{\jmath}} - (T_{\mathcal{B}}(\phi'))_{\gamma,\bar{\imath}} \frac{(T_{\mathcal{B}}(\phi'))_{j,\bar{\jmath}}}{(T_{\mathcal{B}}(\phi'))_{j,\bar{\imath}}}$	$\left(\left(G(\phi')^{-1}_{\bullet\mathcal{B}}\right)_{\gamma\bullet} - \frac{(T_{\mathcal{B}}(\phi'))_{\gamma,\bar{\imath}}}{(T_{\mathcal{B}}(\phi'))_{j,\bar{\imath}}} \left(G(\phi')^{-1}_{\bullet\mathcal{B}}\right)_{j\bullet}\right) q(\phi', \upsilon)$
$\bar{\imath}$	$\frac{1}{(T_{\mathcal{B}}(\phi'))_{j,\bar{\imath}}}$	$\frac{(T_{\mathcal{B}}(\phi'))_{j,\bar{\jmath}}}{(T_{\mathcal{B}}(\phi'))_{j,\bar{\imath}}}$	$\frac{\left(G(\phi')^{-1}_{\bullet\mathcal{B}}\right)_{j\bullet} q(\phi', \upsilon)}{(T_{\mathcal{B}}(\phi'))_{j,\bar{\imath}}}$

Finally, $T_{\mathcal{B}'}(\phi')$ is obtained by pivoting on $(T^*)_{i,\bar{\jmath}}$. Observe the following elements of interest from the RHS of $T_{\mathcal{B}'}(\phi')$:

$\bar{\jmath}$	$\frac{\left(G(\phi')^{-1}_{\bullet\mathcal{B}}\right)_{i\bullet} q(\phi', \upsilon)}{(T_{\mathcal{B}}(\phi'))_{i,\bar{\jmath}}}$
$\bar{\imath}$	$\frac{1}{(T_{\mathcal{B}}(\phi'))_{i,\bar{\jmath}} (T_{\mathcal{B}}(\phi'))_{j,\bar{\imath}}} \left((T_{\mathcal{B}}(\phi'))_{i,\bar{\jmath}} \left(G(\phi')^{-1}_{\bullet\mathcal{B}}\right)_{j\bullet} - (T_{\mathcal{B}}(\phi'))_{j,\bar{\jmath}} \left(G(\phi')^{-1}_{\bullet\mathcal{B}}\right)_{i\bullet}\right) q(\phi', \upsilon)$

(4.13)

As the above values match those claimed in the statement of the lemma, this concludes the proof. □

Proposition 4.7 *Let $\mathcal{B}$ be a f.c.b. such that $\mathcal{IR}_{\mathcal{B}}$ is full dimensional, and let $\hbar^i_{\mathcal{B}}$ be a $(k-1)$-dimensional boundary of $\mathcal{IR}_{\mathcal{B}}$. For any f.c.b. $\mathcal{B}' \neq \mathcal{B}$ such that $|\mathcal{B} \cap \mathcal{B}'| \geq h-2$, $\mathcal{IR}_{\mathcal{B}}$ and $\mathcal{IR}_{\mathcal{B}'}$ are adjacent along $\hbar^i_{\mathcal{B}}$ if and only if one of the following conditions holds:*

1. $\mathcal{B}' = (\mathcal{B} \setminus \{i\}) \cup \{\bar{\imath}\}$ *and* $(T_{\mathcal{B}}(\phi))_{i,\bar{\imath}} \not\equiv 0$.
2. $\mathcal{B}' = (\mathcal{B} \setminus \{i, j\}) \cup \{\bar{\imath}, \bar{\jmath}\}$, $(T_{\mathcal{B}}(\phi))_{i,\bar{\imath}} \equiv 0$, *and the following NLP has a strictly positive optimal value:*

$$NLP_A(\mathcal{B}, i, j) :=$$

$$\begin{aligned}
\max_{\lambda,\phi,\upsilon} \quad & \lambda \\
s.t. \quad & g_{\mathcal{B}}\left(Adj(G(\phi)_{\bullet\mathcal{B}})\right)_{j\bullet} G(\phi)_{\bullet\bar{\imath}} \geq \lambda \\
& g_{\mathcal{B}}\left(Adj(G(\phi)_{\bullet\mathcal{B}})\right)_{\xi\bullet} q(\phi,\upsilon) \geq \lambda \quad \forall \xi \in \left(\mathcal{B} \setminus \left(Z_{\mathcal{B}} \cup H^{i}_{\mathcal{B}} \cup \{i\}\right)\right) \\
& \left(Adj(G(\phi)_{\bullet\mathcal{B}})\right)_{i\bullet} q(\phi,\upsilon) = 0 \\
& g_{\mathcal{B}'}\left(Adj(G(\phi)_{\bullet\mathcal{B}'})\right)_{\xi\bullet} q(\phi,\upsilon) \geq \lambda \quad \forall \xi \in \left(\mathcal{B}' \setminus \left(Z_{\mathcal{B}'} \cup H^{\bar{\jmath}}_{\mathcal{B}'} \cup \{\bar{\jmath}\}\right)\right) \\
& \phi \in \Phi, \upsilon \in \Upsilon
\end{aligned} \tag{4.14}$$

Proof ($\Leftarrow$): Consider the two conditions:
Condition 1: $\mathcal{B}' = (\mathcal{B} \setminus \{i\}) \cup \{\bar{\imath}\}$ and $(T_{\mathcal{B}}(\phi))_{i,\bar{\imath}} \not\equiv 0$.

From Observation 4.4, it is clear that $\mathcal{B}'$ is a basis. We also know that $\mathcal{IR}_{\mathcal{B}}$ and $\mathcal{IR}_{\mathcal{B}'}$ are adjacent along $\hbar^{i}_{\mathcal{B}}$ because: (i) $\left(\bigcup_{\upsilon\in\Upsilon} q(\phi,\upsilon) \cap cone\left(G(\phi)^{-1}_{\bullet(\mathcal{B}\setminus\{i\})}\right)\right)$ is a common boundary of both $\mathcal{ID}_{\mathcal{B}}$ and $\mathcal{ID}_{\mathcal{B}'}$, and (ii) $\mathcal{ID}_{\mathcal{B}}$ and $\mathcal{ID}_{\mathcal{B}'}$ can share this common boundary if and only if $\hbar^{i}_{\mathcal{B}}$ is a common boundary of both $\mathcal{IR}_{\mathcal{B}}$ and $\mathcal{IR}_{\mathcal{B}'}$.
Condition 2: $\mathcal{B}' = (\mathcal{B} \setminus \{i, j\}) \cup \{\bar{\imath}, \bar{\jmath}\}$, $(T_{\mathcal{B}}(\phi))_{i,\bar{\imath}} \equiv 0$, and $NLP_A(\mathcal{B}, i, j)$ has a strictly positive optimal value.

Let $(\lambda^*, \phi^*, \upsilon^*)$ be a solution to $NLP_A(\mathcal{B}, i, j)$ for which $\lambda^* > 0$. We will show that the existence of this solution implies that (i) $\mathcal{B}'$ is a basis, (ii) $\hbar^{i}_{\mathcal{B}}$ forms a $(k-1)$-dimensional boundary of both $\mathcal{IR}_{\mathcal{B}}$ and $\mathcal{IR}_{\mathcal{B}'}$, and (iii) $dim\left(\left(\hbar^{i}_{\mathcal{B}} \cap \mathcal{IR}_{\mathcal{B}}\right) \cap \left(\hbar^{i}_{\mathcal{B}} \cap \mathcal{IR}_{\mathcal{B}'}\right)\right) = k-1$. First, recognize from the definition of NLP_A that because $(\lambda^*, \phi^*, \upsilon^*)$ is feasible to $NLP_A(\mathcal{B}, i, j)$ and $\lambda^* > 0$, then $g_{\mathcal{B}}\left(Adj(G(\phi^*)_{\bullet\mathcal{B}})\right)_{j\bullet} G(\phi^*)_{\bullet\bar{\imath}} > 0$. However, from (2.5), we know that this can be true if and only if $(T_{\mathcal{B}}(\phi^*))_{j,\bar{\imath}} = \dfrac{(Adj(G(\phi^*)_{\bullet\mathcal{B}}))_{j\bullet} G(\phi^*)_{\bullet\bar{\imath}}}{det\left(G(\phi^*)_{\bullet\mathcal{B}}\right)} > 0$. Thus, since the defining constraints of NLP_A ensure that $\phi^* \in \mathrm{Proj}_U \mathcal{IR}_{\mathcal{B}}$, we see from Proposition 4.4 that $\mathcal{B}'$ must be a basis.

Now, recognize that since $(\lambda^*, \phi^*, \upsilon^*)$ is feasible to NLP_A and $\lambda^* > 0$, there must exist $\epsilon > 0$ such that for all $(\phi', \upsilon') \in B_\epsilon(\phi^*, \upsilon^*)$ the following hold: (i) $g_{\mathcal{B}}\left(Adj(G(\phi')_{\bullet\mathcal{B}})\right)_{j\bullet} G(\phi')_{\bullet\bar{\imath}} > 0$, (ii) $g_{\mathcal{B}}\left(Adj(G(\phi')_{\bullet\mathcal{B}})\right)_{\xi\bullet} q(\phi', \upsilon') > 0$ for all $\xi \in \left(\mathcal{B} \setminus \left(\mathrm{Z}_{\mathcal{B}} \cup \mathrm{H}^{i}_{\mathcal{B}} \cup \{i\}\right)\right)$, and (iii) $g_{\mathcal{B}'}\left(Adj(G(\phi')_{\bullet\mathcal{B}'})\right)_{\xi\bullet} q(\phi', \upsilon') > 0$ for all $\xi \in \left(\mathcal{B}' \setminus \left(\mathrm{Z}_{\mathcal{B}'} \cup \mathrm{H}^{\bar{\jmath}}_{\mathcal{B}'} \cup \{\bar{\jmath}\}\right)\right)$. Consider first item (i) and recognize from (4.8) that item (i) is true if and only if $(T_{\mathcal{B}}(\phi'))_{j,\bar{\imath}} > 0$ for all $(\phi', \upsilon') \in B_\epsilon(\phi^*, \upsilon^*)$. This result, together with Lemma 4.3, shows that item (i) is true if and only if $(T_{\mathcal{B}}(\phi'))_{\bar{\imath},j} < 0$ for all $(\phi', \upsilon') \in B_\epsilon(\phi^*, \upsilon^*)$. Next, observe from Lemma 4.4 that two equivalent representations of the RHS of $T_{\mathcal{B}'}(\phi)$ associated with index $\bar{\jmath}$ are $\left(G(\phi)^{-1}_{\bullet\mathcal{B}'}\right)_{\bar{\jmath}\bullet} q(\phi, \upsilon) = \dfrac{\left(Adj(G(\phi)_{\bullet\mathcal{B}'})\right)_{\bar{\jmath}\bullet} q(\phi, \upsilon)}{det\left(G(\phi)_{\bullet\mathcal{B}'}\right)}$

and $\frac{\left(G(\phi)_{\bullet\mathcal{B}}^{-1}\right)_{i\bullet} q(\phi,\upsilon)}{(T_{\mathcal{B}}(\phi))_{i,\bar{J}}} = \frac{(Adj(G(\phi)_{\bullet\mathcal{B}}))_{i\bullet} q(\phi,\upsilon)}{det\,(G(\phi)_{\bullet\mathcal{B}})\,(T_{\mathcal{B}}(\phi))_{i,\bar{J}}}$. Thus, since $det\,(G(\phi)_{\bullet\mathcal{B}})$ is nonzero for all $\phi \in \text{Proj}_U\,\mathcal{IR}_{\mathcal{B}}$ and $(T_{\mathcal{B}}(\phi'))_{\bar{\imath},j} < 0$ for all $(\phi',\upsilon') \in B_\epsilon(\phi^*,\upsilon^*)$, we see from (4.1) that the equivalence of these two representations for the RHS of $T_{\mathcal{B}'}(\phi)$ associated with index $\bar{J}$ implies that $\left(\hbar^i_{\mathcal{B}} \cap \mathcal{IR}_{\mathcal{B}} \cap B_\epsilon(\phi^*,\upsilon^*)\right) = \left(\hbar^{\bar{J}}_{\mathcal{B}'} \cap \mathcal{IR}_{\mathcal{B}} \cap B_\epsilon(\phi^*,\upsilon^*)\right)$. Furthermore, this result and (4.6) show that for every $\ell \in \mathrm{H}^{\bar{J}}_{\mathcal{B}'}$, $(\hbar^i_{\mathcal{B}} \cap \mathcal{IR}_{\mathcal{B}} \cap B_\epsilon(\phi^*,\upsilon^*)) \subseteq (\hbar^\ell_{\mathcal{B}'} \cap \mathcal{IR}_{\mathcal{B}} \cap B_\epsilon(\phi^*,\upsilon^*))$. Now recognize that the equality constraint of NLP_A and the feasibility of $\theta^* = (\phi^*,\upsilon^*)$ to NLP_A show that $\theta^* \in \hbar^i_{\mathcal{B}} \cap \mathcal{IR}_{\mathcal{B}} \cap \mathcal{IR}_{\mathcal{B}'}$. Moreover, the arguments made so far, together with items (ii) and (iii) above, show that θ^* is a point at which all defining constraints of both $\mathcal{IR}_{\mathcal{B}}$ and $\mathcal{IR}_{\mathcal{B}'}$ implied by $\hbar^i_{\mathcal{B}}$ are satisfied at equality and all other defining constraints of both invariancy regions are satisfied strictly. Recognize that such a point can only exist if $\hbar^i_{\mathcal{B}}$ forms a $(k-1)$-dimensional boundary of both $\mathcal{IR}_{\mathcal{B}}$ and $\mathcal{IR}_{\mathcal{B}'}$ and, thus, $dim\left(\left(\hbar^i_{\mathcal{B}} \cap \mathcal{IR}_{\mathcal{B}}\right) \cap \left(\hbar^i_{\mathcal{B}} \cap \mathcal{IR}_{\mathcal{B}'}\right)\right) = k-1$.

($\Rightarrow$): Since $\mathcal{B}'$ is a f.c.b. such that $\mathcal{B}' \neq \mathcal{B}$, $|\mathcal{B} \cap \mathcal{B}'| \geq h-2$, and $\mathcal{IR}_{\mathcal{B}}$ and $\mathcal{IR}_{\mathcal{B}'}$ are adjacent along $\hbar^i_{\mathcal{B}}$, one of the following two cases must occur:

Case 1: $|\mathcal{B} \cap \mathcal{B}'| = h-1$

Here the adjacency of $\mathcal{IR}_{\mathcal{B}}$ and $\mathcal{IR}_{\mathcal{B}'}$ clearly shows that $\mathcal{B}' = (\mathcal{B} \setminus \{i\}) \cup \{\bar{\imath}\}$ is a basis. Hence, by Observation 4.4, there exists $\phi \in \Phi$ such that $(T_{\mathcal{B}}(\phi))_{i,\bar{\imath}} \neq 0$ and Condition 1 is therefore satisfied.

Case 2: $|\mathcal{B} \cap \mathcal{B}'| = h-2$

In this case, $\mathcal{B}' = (\mathcal{B} \setminus \{i\}) \cup \{\bar{\imath}\}$ cannot be a basis and therefore there cannot exist $\phi \in \Phi$ such that $(T_{\mathcal{B}}(\phi))_{i,\bar{\imath}} \neq 0$, i.e., $(T_{\mathcal{B}}(\phi))_{i,\bar{\imath}}$ is identically zero. Furthermore, the fact that $\mathcal{IR}_{\mathcal{B}}$ and $\mathcal{IR}_{\mathcal{B}'}$ are adjacent along $\hbar^i_{\mathcal{B}}$ implies that there exists $j \in \mathcal{B}$ such that $\mathcal{B}' = (\mathcal{B} \setminus \{i,j\}) \cup \{\bar{\imath},\bar{J}\}$ is a basis. Now, since $\mathcal{IR}_{\mathcal{B}}$ and $\mathcal{IR}_{\mathcal{B}'}$ are adjacent across $\hbar^i_{\mathcal{B}}$, we know $dim\left(\left(\hbar^i_{\mathcal{B}} \cap \mathcal{IR}_{\mathcal{B}}\right) \cap \left(\hbar^i_{\mathcal{B}} \cap \mathcal{IR}_{\mathcal{B}'}\right)\right) = k-1$. Furthermore, recognize that we can only have $dim\left(\left(\hbar^i_{\mathcal{B}} \cap \mathcal{IR}_{\mathcal{B}}\right) \cap \left(\hbar^i_{\mathcal{B}} \cap \mathcal{IR}_{\mathcal{B}'}\right)\right) = k-1$ if there exists a point $\theta' = (\phi',\upsilon')$ that satisfies the following: (i) $\theta' \in \left(\hbar^i_{\mathcal{B}} \cap \mathcal{IR}_{\mathcal{B}}\right) \cap \left(\hbar^i_{\mathcal{B}} \cap \mathcal{IR}_{\mathcal{B}'}\right)$, (ii) all defining constraints of $\mathcal{IR}_{\mathcal{B}}$ that are not identically zero and not implied by $\hbar^i_{\mathcal{B}}$ are satisfied strictly at θ', and (iii) all defining constraints of $\mathcal{IR}_{\mathcal{B}'}$ that are not identically zero and not implied by $\hbar^i_{\mathcal{B}}$ are satisfied strictly at θ'. First, notice that item (i) implies that

$$\left(Adj(G(\phi')_{\bullet\mathcal{B}})\right)_{i\bullet} q(\phi',\upsilon') = 0. \tag{4.15}$$

Next, note that the combination of items (i), (ii), and (iii) above implies that $q(\phi',\upsilon')$ lies in both $\mathcal{C}_{\mathcal{B}'}(\phi')$ and the relative interior of $cone\left(G(\phi')_{\bullet(\mathcal{B}\setminus\{i\})}\right)$. Hence, from Corollary 4.2, we have $(T_{\mathcal{B}}(\phi'))_{j,\bar{\imath}} = \dfrac{\left(Adj(G(\phi')_{\cdot\mathcal{B}})\right)_{j\cdot} G(\phi')_{\cdot\bar{\imath}}}{det(G(\phi')_{\cdot\mathcal{B}})} > 0$, or equivalently,

$$g_{\mathcal{B}}\left(Adj(G(\phi')_{\cdot\mathcal{B}})\right)_{j\cdot} G(\phi')_{\cdot\bar{\imath}} > 0. \tag{4.16}$$

Additionally, note that because the defining constraints of $\mathcal{IR}_{\mathcal{B}}$ that are identically zero are those whose associated indices are in $\mathrm{Z}_{\mathcal{B}}$ and the defining constraints of $\mathcal{IR}_{\mathcal{B}}$ that are implied by $\hbar^i_{\mathcal{B}}$ are those whose associated indices are in $\mathrm{H}^i_{\mathcal{B}}$, item (ii) above implies that

$$g_{\mathcal{B}}\left(Adj(G(\phi')_{\cdot\mathcal{B}})\right)_{\xi\cdot} q(\phi', \upsilon') > 0 \text{ for all } \xi \in \left(\mathcal{B} \setminus \left(\mathrm{Z}_{\mathcal{B}} \cup \mathrm{H}^i_{\mathcal{B}} \cup \{i\}\right)\right). \tag{4.17}$$

Now, notice that because $(T_{\mathcal{B}}(\phi'))_{j,\bar{\imath}} > 0$, there must exist $\epsilon > 0$ such that for all $\theta^* \in B_\epsilon(\theta')$, $(T_{\mathcal{B}}(\phi^*))_{j,\bar{\imath}} > 0$. Then by the same arguments used in the proof of the reverse direction, we have $\left(\hbar^i_{\mathcal{B}} \cap \mathcal{IR}_{\mathcal{B}} \cap B_\epsilon(\theta')\right) = \left(\hbar^{\bar{J}}_{\mathcal{B}'} \cap \mathcal{IR}_{\mathcal{B}} \cap B_\epsilon(\theta')\right)$, i.e., within the ϵ-ball centered at θ', the sets $\hbar^i_{\mathcal{B}}$ and $\hbar^{\bar{J}}_{\mathcal{B}'}$ are identical. This shows that item (iii) above can be equivalently stated as "all defining constraints of $\mathcal{IR}_{\mathcal{B}'}$ that are not identically zero and not implied by $\hbar^{\bar{J}}_{\mathcal{B}'}$ are satisfied strictly at θ'." Therefore, because the defining constraints of $\mathcal{IR}_{\mathcal{B}'}$ that are identically zero are those whose associated indices are in $\mathrm{Z}_{\mathcal{B}'}$ and the defining constraints of $\mathcal{IR}_{\mathcal{B}'}$ that are implied by $\hbar^{\bar{J}}_{\mathcal{B}'}$ are those whose associated indices are in $\mathrm{H}^{\bar{J}}_{\mathcal{B}'}$, item (iii) implies that

$$g_{\mathcal{B}'}\left(Adj(G(\phi')_{\cdot\mathcal{B}'})\right)_{\xi\cdot} q(\phi', \upsilon') > 0 \text{ for all } \xi \in \left(\mathcal{B}' \setminus \left(\mathrm{Z}_{\mathcal{B}'} \cup \mathrm{H}^{\bar{J}}_{\mathcal{B}'} \cup \{\bar{J}\}\right)\right). \tag{4.18}$$

Define λ' to be the minimum of the set of values containing the LHS value of the inequality displayed in (4.16) and the LHS values of each of the inequalities displayed in (4.17) and (4.18), as given by the specified values of ξ. It is clear that $\lambda' > 0$. Also, the combination of (4.15), (4.16), (4.17), and (4.18) show that $(\lambda', \phi', \upsilon')$ is feasible to NLP_A. Therefore, the optimal value of NLP_A must be strictly positive. □

Corollary 4.3 *Let a f.c.b. $\mathcal{B}$ be given, together with distinct $i, j \in \mathcal{B}$. Consider $\mathcal{B}' = (\mathcal{B} \setminus \{i, j\}) \cup \{\bar{\imath}, \bar{J}\}$. If $(T_{\mathcal{B}}(\phi))_{j,\bar{\imath}}$ is a nonpositive constant, then $NLP_A(\mathcal{B}, i, j)$ cannot have a strictly positive optimal value.*

Proof Recognize that $(T_{\mathcal{B}}(\phi))_{j,\bar{\imath}} = \dfrac{(Adj(G(\phi)_{\cdot\mathcal{B}}))_{j\cdot} G(\phi)_{\cdot\bar{\imath}}}{det\,(G(\phi)_{\cdot\mathcal{B}})}$. Thus, $(T_{\mathcal{B}}(\phi))_{j,\bar{\imath}}$ is nonpositive if and only if $\dfrac{(Adj(G(\phi)_{\cdot\mathcal{B}}))_{j\cdot} G(\phi)_{\cdot\bar{\imath}}}{det\,(G(\phi)_{\cdot\mathcal{B}})}$ is nonpositive. Furthermore, recall from (2.5) that $g_{\mathcal{B}}$ has the same sign as $det\,(G(\phi)_{\cdot\mathcal{B}})$ for all $\phi \in \Phi$ such that $det\,(G(\phi)_{\cdot\mathcal{B}}) \neq 0$. Thus, $(T_{\mathcal{B}}(\phi))_{j,\bar{\imath}}$ is nonpositive if and only if

$g_{\mathcal{B}}(Adj(G(\phi)_{\bullet\mathcal{B}}))_{j\bullet}G(\phi)_{\bullet\bar{\imath}}$ is nonpositive. Therefore, if $(T_{\mathcal{B}}(\phi))_{j,\bar{\imath}}$ is nonpositive, $NLP_A(\mathcal{B}, i, j)$ cannot have a strictly positive optimal value due to its first constraint. □

Notice that condition 1 of Proposition 4.7 indicates situations in which diagonal pivots can be used to obtain new adjacent invariancy regions, while condition 2 indicates the situations in which exchange pivots can be used to obtain new adjacent invariancy regions. Hence, combining Lemma 4.4, Proposition 4.7, and Corollary 4.3 leads us to a strategy so that, given a full dimensional invariancy region associated with a f.c.b. $\mathcal{B}$ and an $i \in \mathcal{B}$ such that $h^{i}_{\mathcal{B}}$ forms a $(k-1)$-dimensional boundary of $\mathcal{IR}_{\mathcal{B}}$, we can compute the set of all f.c.b.'s whose invariancy regions are adjacent to $\mathcal{IR}_{\mathcal{B}}$ across $h^{i}_{\mathcal{B}}$. This procedure is outlined in Algorithm 4.

On lines 1–5 of Algorithm 4, condition 1 of Proposition 4.7 is used to attempt to find adjacent regions using diagonal pivots, and on lines 6–12, condition 2 of Proposition 4.7 is used to attempt to find adjacent regions using exchange pivots. Note that the result of Corollary 4.3 is exploited in the condition of line 9. Additionally, notice that when a f.c.b. is discovered whose associated invariancy region is adjacent to the current invariancy region, there is no guarantee that the new region is full dimensional. Hence, Algorithm 4 relies on two subroutines, GETINVARIANCYREGIONDIMENSION and FINDADJACENTKMINUS1, discussed in detail in Sections 4.2.4 and 4.2.5, respectively. The routine GETINVARIANCYREGIONDIMENSION is used to ascertain the dimension of a newly discovered region, while FINDADJACENTKMINUS1 is used to find full dimensional regions adjacent

---

Algorithm 4 GETADJACENTREGIONSACROSS($\mathcal{B}$,i,$\mathscr{B}$)—Determine all f.c.b.'s with associated invariancy regions that are adjacent to $\mathcal{IR}_{\mathcal{B}}$ across $h^{i}_{\mathcal{B}}$

Input: A f.c.b. $\mathcal{B}$, an index $i \in \mathrm{F}_{\mathcal{B}}$, and the set $\mathscr{B}$ of previously processed f.c.b.'s.
Output: The set $\mathcal{S}'$ of f.c.b.'s with associated invariancy regions that are adjacent to $\mathcal{IR}_{\mathcal{B}}$ across $h^{i}_{\mathcal{B}}$ and an updated version of $\mathscr{B}$.

1: Set $\mathcal{S}' = \emptyset$ and $\mathcal{B}^* = \big(\mathcal{B} \setminus \{i\}\big) \cup \{\bar{\imath}\}$.
2: **if** $\mathcal{B}^* \notin \mathscr{B}$ **then**
3:     **if** $(T_{\mathcal{B}}(\phi))_{i,\bar{\imath}} \not\equiv 0$ **then** set $\mathscr{B} = \mathscr{B} \cup \{\mathcal{B}^*\}$ and let $d =$ GETINVARIANCYREGIONDIMENSION($\mathcal{B}^*$).
4:         **if** $d = k$ **then** set $\mathcal{S}' = \{\mathcal{B}^*\}$.
5:         **else** set $(\mathcal{S}', \mathscr{B}) =$ FINDADJACENTKMINUS1($\mathcal{B}^*, \bar{\imath}, \mathscr{B}$).
6:     **else**
7:         **for** $j \in \mathcal{B} \setminus \{i\}$ **do**
8:             Set $\mathcal{B}^* = \big(\mathcal{B} \setminus \{i, j\}\big) \cup \{\bar{\imath}, \bar{\jmath}\}$
9:             **if** $\mathcal{B}^* \notin \mathscr{B}$ and $(T_{\mathcal{B}}(\phi))_{j,\bar{\imath}}$ either has degree of at least 1 or is a strictly positive constant **then** solve $NLP_A(\mathcal{B}, i, j)$ to obtain an optimal solution $(\lambda^*, \phi^*, \upsilon^*)$.
10:                 **if** $\lambda^* > 0$ **then** set $\mathscr{B} = \mathscr{B} \cup \{\mathcal{B}^*\}$ and let $d =$ GETINVARIANCYREGIONDIMENSION($\mathcal{B}^*$).
11:                     **if** $d = k$ **then** set $\mathcal{S}' = \mathcal{S}' \cup \{\mathcal{B}^*\}$.
12:                     **else** set $(\mathcal{S}'', \mathscr{B}) =$ FINDADJACENTKMINUS1($\mathcal{B}^*, \bar{\jmath}, \mathscr{B}$) and $\mathcal{S}' = \mathcal{S}'' \cup \mathcal{S}'$.
13: Return $(\mathcal{S}', \mathscr{B})$.

---

to a given $(k-1)$-dimensional region. We note here that the second argument of the routine FINDADJACENTKMINUS1 must be an index, say ℓ, in the f.c.b. passed as the routine's first argument, say $\mathcal{B}'$, such that $dim\left(\mathcal{IR}_{\mathcal{B}'} \cap \hbar^{\ell}_{\mathcal{B}'}\right) = k-1$. We claim here that the choice of $\bar{i}$ on line 5 and $\bar{j}$ on line 12 satisfies the aforementioned property. Justification for this claim is provided in Proposition 4.14, presented at the end of Section 4.2.5.

We now again consider Examples 2.1 and 2.2. However, in order to highlight certain key qualities present in each example, we now consider them separately, beginning with Example 2.1. Due to the size of the tableaux associated with this example, they cannot be displayed here but are instead included in Appendix A. During our previous consideration of Example 2.1, we found that $\hbar^{w_1}_{\mathcal{B}_0^{2.1}}$ and $\hbar^{z_3}_{\mathcal{B}_0^{2.1}}$ formed $(k-1)$-dimensional boundaries of $\mathcal{IR}_{\mathcal{B}_0^{2.1}}$ and so $\mathrm{F}_{\mathcal{B}_0^{2.1}} = \{w_1, z_3\}$. Hence, in continuing to process $\mathcal{B}_0^{2.1}$, we return to Algorithm 1 and enter the for loop on line 4. The first call is $(\mathcal{S}', \mathscr{B}) =$ GETADJACENTREGIONSACROSS$(\mathcal{B}_0^{2.1}, w_1, \mathscr{B})$. Note that currently $\mathscr{B} = \left\{\mathcal{B}_0^{2.1}\right\}$. Observe from Table A.7 in Appendix A that $(T_{\mathcal{B}_0^{2.1}}(\phi))_{w_1,z_1} = \frac{4\,(32\,\phi_2-7\,\phi_1+29)}{-3\,\phi_1^2+8\,\phi_1\,\phi_2+19\,\phi_1+41\,\phi_2^2-24\,\phi_2-22} \not\equiv 0$. Hence, by condition 1 of Proposition 4.7, a diagonal pivot on this element yields f.c.b. $\mathcal{B}_1^{2.1} = \{z_1, z_2, z_3, z_4, z_5\}$ with $\mathcal{IR}_{\mathcal{B}_1^{2.1}}$ adjacent to $\mathcal{IR}_{\mathcal{B}_0^{2.1}}$ across $\hbar^{w_1}_{\mathcal{B}_0^{2.1}}$. Thus, the condition of the if statement on line 3 of Algorithm 4 is satisfied, and we now call $d =$ GETINVARIANCYREGIONDIMENSION$(\mathcal{B}_1^{2.1})$, which we will further consider in Section 4.2.4.

Let us now return our focus to Example 2.2. During our previous consideration of Example 2.2, we found that $\hbar^{w_2}_{\mathcal{B}_0^{2.2}}$ and $\hbar^{w_3}_{\mathcal{B}_0^{2.2}} = \hbar^{w_4}_{\mathcal{B}_0^{2.2}}$ formed $(k-1)$-dimensional boundaries of $\mathcal{IR}_{\mathcal{B}_0^{2.2}}$ and so $\mathrm{F}_{\mathcal{B}_0^{2.2}} = \{w_2, w_3, w_4\}$. Hence, we return to Algorithm 1 and enter the for loop on line 4. The first call is $(\mathcal{S}', \mathscr{B}) =$ GETADJACENTREGIONSACROSS$(\mathcal{B}_0^{2.2}, w_2, \mathscr{B})$ (currently $\mathscr{B} = \left\{\mathcal{B}_0^{2.2}\right\}$). From Table B.1 in Appendix B, we observe that $(T_{\mathcal{B}_0^{2.2}}(\phi))_{w_2,z_2} \equiv 0$. This shows that condition 1 of Proposition 4.7 does not apply, and so we must use condition 2. Thus, we now enter the for loop on line 7 of Algorithm 4 and check for exchange pivots involving w_2 and either w_1, w_3, or w_4. Observe the following vector, which is the z_2 column of $T_{\mathcal{B}_0^{2.2}(\phi)}$: $\begin{bmatrix} 0 \\ 0 \\ -3 \\ \phi+5 \end{bmatrix}$.

Here we see that the condition of the if statement on line 9 is only satisfied by w_4. Hence, we only consider the exchange pivot involving w_2 and w_4. Observe $NLP_A(\mathcal{B}_0^{2.2}, w_2, w_4)$ (4.14):

$$\begin{aligned}
\max_{\lambda,\phi,\upsilon} \quad & \lambda \\
\text{s.t.} \quad & \phi + 5 \geq \lambda \\
& -\upsilon - 1 \geq \lambda \\
& -18\upsilon - 34 \geq \lambda \\
& -9\upsilon - 17 \geq \lambda \\
& \phi - \upsilon - 1 = 0 \\
& -8\upsilon - \phi\upsilon - 8 \geq \lambda \\
& -(2\phi + 13)(9\upsilon + 17) \geq \lambda \\
& \phi \in [-3, 1], \upsilon \in [-3, 1]
\end{aligned}$$

An optimal solution to $NLP_A(\mathcal{B}_0^{2.2}, w_2, w_4)$ is $(\lambda^*, \phi^*, \upsilon^*) = (2, -2, -3)$, which shows that by condition 2 of Proposition 4.7, the invariancy region associated with f.c.b. $\mathcal{B}_1^{2.2} = \{w_1, z_2, w_3, z_4\}$ is adjacent to $\mathcal{IR}_{\mathcal{B}_0^{2.2}}$ across $\hbar^{w_2}_{\mathcal{B}_0^{2.2}}$. Therefore, the condition of the if statement on line 10 is satisfied, and so we call d = GETINVARIANCYREGIONDIMENSION($\mathcal{B}_1^{2.2}$), which is also discussed further in Section 4.2.4.

As we can proceed no further in our consideration of either Example 2.1 or 2.2 without an understanding of the procedure GETINVARIANCYREGIONDIMENSION, we now advance to Section 4.2.4 where we outline this procedure.

4.2.4 *Identification of Full Dimensional Invariancy Regions*

In the discussion that follows, we consider the procedure GETINVARIANCYREGIONDIMENSION, which is carried out by determining whether or not the invariancy region associated with the f.c.b. passed into the procedure is full dimensional. From our earlier discourse (particularly Proposition 3.11), it is clear that in order to partition Θ it is only necessary to consider invariancy regions of dimension $k - 1$ or greater. Furthermore, we note that the algorithms presented throughout this work are all organized in order to guarantee that any f.c.b. passed to GETINVARIANCYREGIONDIMENSION will have an invariancy region that is at least $(k - 1)$-dimensional (see Corollary 4.4). Hence, whenever the procedure GETINVARIANCYREGIONDIMENSION fails to conclude that an invariancy region is k-dimensional, we can conclude that its dimension is $k - 1$. The following proposition provides the tools necessary for checking this dimension.

Proposition 4.8 *Let a f.c.b. $\mathcal{B}$ be given. Then $\mathcal{IR}_{\mathcal{B}}$ is full dimensional if and only if $|Z_{\mathcal{B}}| \leq h - (k - p)$, and there exist $\phi' \in \Phi$ and $\upsilon' \in \Upsilon$ such that $g_{\mathcal{B}}\left(Adj(G(\phi')_{\cdot\mathcal{B}})\right)_{i.} q(\phi', \upsilon') > 0$ for all $i \in \mathcal{B} \setminus Z_{\mathcal{B}}$, i.e., if the following NLP has a strictly positive optimal value:*

$$
\begin{aligned}
NLP_D(\mathcal{B}) := \max_{\lambda,\phi,\upsilon} \quad & \lambda \\
\text{s.t.} \quad & g_{\mathcal{B}}(Adj(G(\phi)_{\bullet\mathcal{B}}))_{i\bullet} q(\phi,\upsilon) \ge \lambda \qquad \forall i \in \mathcal{B} \setminus Z_{\mathcal{B}} \\
& \phi \in \Phi, \ \upsilon \in \Upsilon
\end{aligned}
\tag{4.19}
$$

Proof (⇐): Let $(\lambda^*, \phi^*, \upsilon^*)$ be an optimal solution to $NLP_D(\mathcal{B})$ and $\lambda^* > 0$. Because $g_{\mathcal{B}}\left(Adj(G(\phi')_{\bullet\mathcal{B}})\right)_{i\bullet} q(\phi', \upsilon') > 0$ for all $i \in \mathcal{B} \setminus Z_{\mathcal{B}}$, $\bigcup_{\upsilon\in\Upsilon} q(\phi^*, \upsilon)$ must intersect the relative interior of $cone\left(G(\phi^*)_{\bullet(\mathcal{B}\setminus Z_{\mathcal{B}})}\right)$. Furthermore, since $\bigcup_{\upsilon\in\Upsilon} q(\phi^*, \upsilon)$ intersects the relative interior of $cone\left(G(\phi^*)_{\bullet(\mathcal{B}\setminus Z_{\mathcal{B}})}\right)$, there must exist $\epsilon > 0$ such that $\bigcup_{\upsilon\in\Upsilon} q(\phi, \upsilon)$ intersects the relative interior of $cone\left(G(\phi)_{\bullet(\mathcal{B}\setminus Z_{\mathcal{B}})}\right)$ for all $\phi \in B_\epsilon(\phi^*)$. Thus, since $dim\left(\bigcup_{\upsilon\in\Upsilon} q(\phi, \upsilon)\right) = k - p$ for all $\phi \in \Phi$, we have $dim\left(\bigcup_{\upsilon\in\Upsilon} q(\phi, \upsilon) \cap \mathcal{C}_{\mathcal{B}}(\phi)\right) = \min\{k - p, h - |Z_{\mathcal{B}}|\}$ for all $\phi \in B_\epsilon(\phi^*)$. By Corollary 3.2, this shows that $dim(\mathcal{IR}_{\mathcal{B}}) = k$ since $|Z_{\mathcal{B}}| \le h - (k - p)$.

(⇒): We use contradiction for the proof in this direction. Consider two cases:

Case 1: $|Z_{\mathcal{B}}| > h - (k - p)$

Recognize that for each $i \in Z_{\mathcal{B}}$, the column $G(\phi)_{\bullet i}$ is not needed for the description of $\bigcup_{\upsilon\in\Upsilon} q(\phi, \upsilon) \cap \mathcal{C}_{\mathcal{B}}(\phi)$ for all $\phi \in \Phi$. So we have $\left(\bigcup_{\upsilon\in\Upsilon} q(\phi, \upsilon) \cap \mathcal{C}_{\mathcal{B}}(\phi)\right) \subseteq cone\left(G(\phi)_{\bullet(\mathcal{B}\setminus Z_{\mathcal{B}})}\right)$ for all $\phi \in \Phi$. However, because $|Z_{\mathcal{B}}| > h - (k - p)$, we also have $dim\left(cone\left(G(\phi)_{\bullet(\mathcal{B}\setminus Z_{\mathcal{B}})}\right)\right) < k - p$ for all $\phi \in \Phi$. This shows that $dim\left(\bigcup_{\upsilon\in\Upsilon} q(\phi, \upsilon) \cap \mathcal{C}_{\mathcal{B}}(\phi)\right) < k - p$ for all $\phi \in \Phi$ and, thus, by Corollary 3.2, $\mathcal{IR}_{\mathcal{B}}$ cannot be full dimensional, which is a contradiction.

Case 2: There does not exist a solution $(\lambda^*, \phi^*, \upsilon^*)$ to $NLP_D(\mathcal{B})$ such that $\lambda^* > 0$

Note that if $NLP_D(\mathcal{B})$ is not feasible, then $\mathcal{IR}_{\mathcal{B}}$ is empty. This is a contradiction, however, to the fact that $\mathcal{B}$ is a f.c.b. Now suppose that for all solutions $(\lambda^*, \phi^*, \upsilon^*)$ to $NLP_D(\mathcal{B})$, $\lambda^* \le 0$. In this case, for all $\theta \in \mathcal{IR}_{\mathcal{B}}$ and all $\epsilon > 0$, we have $B_\epsilon(\theta) \not\subset \mathcal{IR}_{\mathcal{B}}$. Therefore, $\mathcal{IR}_{\mathcal{B}}$ cannot be full dimensional, which is a contradiction.

The contradictions found in the cases above show that there must be a solution $(\lambda^*, \phi^*, \upsilon^*)$ for $NLP_D(\mathcal{B})$ such that $\lambda^* > 0$ and $|Z_{\mathcal{B}}| \le h - (k - p)$. □

The result of Proposition 4.8 shows that, for any f.c.b. $\mathcal{B}$, $NLP_D(\mathcal{B})$ can be used to determine whether or not $\mathcal{IR}_{\mathcal{B}}$ is full dimensional. Hence, we present Algorithm 5 that outlines the procedure GETINVARIANCYREGIONDIMENSION.

Note from lines 2 and 3 of Algorithm 5 that $\mathcal{B}$ is identified as having a k-dimensional invariancy region if the optimal value of NLP_D is positive. Otherwise, $\mathcal{B}$ is reported as having a $(k-1)$-dimensional invariancy region. To see that this is justified, recall that in Algorithm 4, the subroutine GETINVARIANCYREGIONDIMENSION is only executed after obtaining a f.c.b. satisfying one of the conditions of Proposition 4.7. The following corollary of Propositions 4.7 and 4.8 ensures that any f.c.b. passed to the subroutine GETINVARIANCYREGIONDIMENSION will have a dimension of at least $k-1$. Thus, combining the results of Proposition 4.8 and Corollary 4.4 guarantees the correctness of Algorithm 5.

Corollary 4.4 *Let distinct f.c.b.'s $\mathcal{B}'$ and $\mathcal{B}$ be given that satisfy: (i) $dim(\mathcal{IR}_{\mathcal{B}}) = k$, (ii) $|\mathcal{B}' \cap \mathcal{B}| \geq h-2$, and (iii) $\mathcal{IR}_{\mathcal{B}}$ is adjacent to $\mathcal{IR}_{\mathcal{B}'}$ along $\mathcal{h}^i_{\mathcal{B}}$. Then $dim\left(\mathcal{IR}_{\mathcal{B}'}\right) \geq k-1$. Furthermore, $dim\left(\mathcal{IR}_{\mathcal{B}'}\right) = k-1$ if and only if $|Z_{\mathcal{B}'}| > h-(k-p)$ or the optimal value of $NLP_D(\mathcal{B}')$ is nonpositive.*

Proof First notice that $dim\left(\mathcal{IR}_{\mathcal{B}'}\right) \geq k-1$ since $\mathcal{h}^i_{\mathcal{B}}$ forms a $(k-1)$-dimensional boundary of both $\mathcal{IR}_{\mathcal{B}}$ and $\mathcal{IR}_{\mathcal{B}'}$, as shown in the proof of Proposition 4.7. The fact that $dim\left(\mathcal{IR}_{\mathcal{B}'}\right) = k-1$ if and only if $|Z_{\mathcal{B}'}| > h-(k-p)$ or the optimal value of $NLP_D(\mathcal{B}')$ is nonpositive follows directly from Proposition 4.8. □

Recall that in Section 4.2.3 we paused our consideration of Examples 2.1 and 2.2 since we were not yet ready to execute GETINVARIANCYREGIONDIMENSION($\mathcal{B}_1^{2.1}$) or GETINVARIANCYREGIONDIMENSION($\mathcal{B}_1^{2.2}$). We resume this discussion now.

Observe $NLP_D(\mathcal{B}_1^{2.1})$ and $NLP_D(\mathcal{B}_1^{2.2})$, respectively, derived from Table A.10 in Appendix A and Table B.2 in Appendix B.

$NLP_D(\mathcal{B}_1^{2.1})$:

Algorithm 5 GETINVARIANCYREGIONDIMENSION($\mathcal{B}$)—Find the dimension of $\mathcal{IR}_{\mathcal{B}}$

Input: A f.c.b. $\mathcal{B}$.
Output: The dimension d of $\mathcal{IR}_{\mathcal{B}}$.

1: Solve $NLP_D(\mathcal{B})$ to obtain an optimal solution $(\lambda^*, \phi^*, \upsilon^*)$.
2: **if** $\lambda^* > 0$ **then** return k.
3: **else** return $k-1$.

$$
\begin{aligned}
\max_{\lambda,\phi} \quad & \lambda \\
\text{s.t.} \quad & 2(3\phi_1^3 + 18\phi_1^2\phi_2 - 49\phi_1^2 - 75\phi_1\phi_2^2 + 148\phi_1\phi_2 + 68\phi_1 + 96\phi_2^2 - 16\phi_2 - 76) \geq \lambda \\
& -81\phi_1^3 + 288\phi_1^2\phi_2 - 111\phi_1^2 + 47\phi_1\phi_2^2 + 856\phi_1\phi_2 + 422\phi_1 + 544\phi_2^2 + 392\phi_2 - 44 \geq \lambda \\
& -4(61\phi_1 - 60\phi_1\phi_2 + 29\phi_1^2 - 12) \geq \lambda \\
& 4(75\phi_1 + 32\phi_2 - 5\phi_1\phi_2 + 13\phi_1^2 + 20) \geq \lambda \\
& 4(16\phi_1 + 96\phi_2 + 9\phi_1\phi_2 + 2\phi_1^2 + 84) \geq \lambda \\
& \phi_1 + \phi_2 \leq 1 \\
& \phi_1, \phi_2 \geq 0
\end{aligned}
$$

$NLP_D(\mathcal{B}_1^{2.2})$:

$$
\begin{aligned}
\max_{\lambda,\phi,\upsilon} \quad & \lambda \\
\text{s.t.} \quad & -(\phi + 5)(8\upsilon + \phi\upsilon + 8) \geq \lambda \\
& -(\phi + 7)(9\upsilon + 17) \geq \lambda \\
& -(\phi + 7)(2\phi + 13)(9\upsilon + 17) \geq \lambda \\
& (\phi + 5)(\upsilon - \phi + 1) \geq \lambda \\
& \phi \in [-3, 1], \upsilon \in [-3, 1]
\end{aligned}
$$

Respective optimal solutions of $NLP_D(\mathcal{B}_1^{2.1})$ and $NLP_D(\mathcal{B}_1^{2.2})$ are approximately $(\lambda^*, \phi_1^*, \phi_2^*) = (8, 0, 1)$ and $(\lambda^{**}, \phi^{**}, \upsilon^{**}) = (4.0007, -2.9718, -1.9992)$. Hence, $\mathcal{IR}_{\mathcal{B}_1^{2.1}}$ and $\mathcal{IR}_{\mathcal{B}_1^{2.2}}$ are both full dimensional.

We now cease consideration of Example 2.1 for good because there are no novel steps. Further examination only yields that $\mathcal{B}_2^{2.1} = \{w_1, z_2, w_3, z_4, z_5\}$ is a f.c.b. with a full dimensional invariancy region and that $\mathcal{IR}_{\mathcal{B}_0^{2.1}}$, $\mathcal{IR}_{\mathcal{B}_1^{2.1}}$, and $\mathcal{IR}_{\mathcal{B}_2^{2.1}}$ partition Θ.

As for Example 2.2, since $\mathcal{IR}_{\mathcal{B}_1^{2.2}}$ is full dimensional, execution of GETADJACENTREGIONSACROSS($\mathcal{B}_0^{2.2}, w_2, \mathscr{B}$) ceases with the return of $(\mathcal{S}' = \{\mathcal{B}_1^{2.2}\}, \mathscr{B} = \{\mathcal{B}_0^{2.2}, \mathcal{B}_1^{2.2}\})$. Hence, we now return to the for loop on line 4 of Algorithm 1 and consider w_3 (recall that $\mathrm{F}_{\mathcal{B}_0^{2.2}} = \{w_2, w_3, w_4\}$ and we have already considered w_2). Thus, the next call is $(\mathcal{S}', \mathscr{B}) =$ GETADJACENTREGIONSACROSS($\mathcal{B}_0^{2.2}, w_3, \mathscr{B}$), and so we move to Algorithm 4. It can be observed from $T_{\mathcal{B}_0^{2.2}}(\phi)$, found in Table B.1 in Appendix B, that due to the if statement on line 9, the only exchange pivots we must now consider are those involving: (i) w_3 and w_1 and (ii) w_3 and w_2. Respective optimal solutions to $NLP_A(\mathcal{B}_0^{2.2}, w_3, w_1)$ and $NLP_A(\mathcal{B}_0^{2.2}, w_3, w_2)$ are approximately $(-0.6667, 1, -1.8889)$ and $(0.8889, -1.62, -1.8889)$. This shows that the basis $\{z_1, w_2, z_3, w_4\}$ does not yield an invariancy region that is adjacent to $\mathcal{IR}_{\mathcal{B}_0^{2.2}}$ across $\hbar^{w_3}_{\mathcal{B}_0^{2.2}}$, but basis $\mathcal{B}_2^{2.2} = \{w_1, z_2, z_3, w_4\}$ does. We therefore call $d =$ GETINVARIANCYREGIONDIMENSION($\mathcal{B}_2^{2.2}$). Observe $NLP_D(\mathcal{B}_2^{2.2})$:

$$\begin{aligned} \max_{\lambda,\phi,\upsilon} \quad & \lambda \\ \text{s.t.} \quad & -3(2+3\,\upsilon+3) \geq \lambda \\ & 90\,\upsilon + 170 \geq \lambda \\ & 3(\phi - \upsilon - 1) \geq \lambda \\ & -5\,(2\,\phi + 13)\;(9\,\upsilon + 17) \geq \lambda \\ & \phi \in [-3, 1], \upsilon \in [-3, 1] \end{aligned}$$

An optimal solution of $NLP_D(\mathcal{B}_2^{2.2})$ is approximately $(\lambda^{***}, \phi^{***}, \upsilon^{***}) = (0, -0.0574, -1.8889)$. Since $\lambda^{***} = 0$, we have that $\mathcal{IR}_{\mathcal{B}_2^{2.2}}$ is $(k-1)$-dimensional. As a result, we must now pause our consideration of Example 2.2 so that we may discuss the details of the procedure FINDADJACENTKMINUS1.

4.2.5 Determining Adjacent Invariancy Regions from a Given $(k-1)$-Dimensional Region

Recall from Section 4.2.4 that Proposition 4.7 provides a strategy for determining the invariancy regions that are adjacent to a given full dimensional invariancy region across a known $(k-1)$-dimensional boundary. Corollary 4.4 provides a strategy for determining when a discovered invariancy region is $(k-1)$-dimensional. Suppose that for some f.c.b. $\mathcal{B}$, we solve $NLP_D(\mathcal{B})$ and discover that $\mathcal{IR}_{\mathcal{B}}$ is not full dimensional. Unfortunately, Corollary 4.4 does not provide any insight into which indices in $\mathcal{B}$ ought to be pivoted on in order to yield a new f.c.b. $\mathcal{B}'$ for which $\mathcal{IR}_{\mathcal{B}'}$ is adjacent to $\mathcal{IR}_{\mathcal{B}}$ and at least $(k-1)$-dimensional. The following discussion addresses finding such indices. Consider the following proposition.

Proposition 4.9 *Let distinct f.c.b.'s $\mathcal{B}'$ and $\mathcal{B}$ be given for which: (i) $dim(\mathcal{IR}_{\mathcal{B}}) = k$, (ii) $|\mathcal{B}' \cap \mathcal{B}| \geq h-2$, and (iii) $\mathcal{IR}_{\mathcal{B}}$ is adjacent to $\mathcal{IR}_{\mathcal{B}'}$ along $\hbar^i_{\mathcal{B}'}$. Then $dim\left(\mathcal{IR}_{\mathcal{B}'}\right) = k-1$ if and only if for every $\theta \in \mathcal{IR}_{\mathcal{B}'}$, there exist an $\epsilon(\theta) > 0$ and an index $j \in \mathcal{B}' \setminus \{i\}$ such that the following hold:*

1. *For all $\theta' = (\phi', \upsilon') \in B_{\epsilon(\theta)}(\theta) \cap \mathcal{IR}_{\mathcal{B}'}$, $\left(Adj(G(\phi')_{\bullet\mathcal{B}'})\right)_{i,\bullet} q(\phi', \upsilon') = \left(Adj(G(\phi')_{\bullet\mathcal{B}'})\right)_{j,\bullet} q(\phi', \upsilon')$.*
2. *$B_{\epsilon(\theta)}(\theta)$ has empty intersection with the following set:* $\{\theta = (\phi, \upsilon) : g_{\mathcal{B}'}\left(Adj(G(\phi)_{\bullet\mathcal{B}'})\right)_{i,\bullet} q(\phi, \upsilon) > 0,\, g_{\mathcal{B}'}\left(Adj(G(\phi)_{\bullet\mathcal{B}'})\right)_{j,\bullet} q(\phi, \upsilon) > 0\}$.

Proof ($\Rightarrow$): Since we know that $\mathcal{IR}_{\mathcal{B}'}$ is $(k-1)$-dimensional and adjacent to another invariancy region across $\hbar^i_{\mathcal{B}'}$, we know $\mathcal{IR}_{\mathcal{B}'} \subset \hbar^i_{\mathcal{B}'}$. Since there are a finite number of constraints defining $\mathcal{IR}_{\mathcal{B}'}$, we know that this can only be the case if for every $\theta \in \mathcal{IR}_{\mathcal{B}'}$, there exists $j \in \mathcal{B}' \setminus \{i\}$ such that: (i) $\hbar^i_{\mathcal{B}'}$

and $h^j_{\mathcal{B}'}$ intersect at θ, (ii) $dim(h^i_{\mathcal{B}'} \cap h^j_{\mathcal{B}'} \cap \mathcal{IR}_{\mathcal{B}'}) = k-1$, and (iii) there exists $\epsilon(\theta) > 0$ such that the intersections of the open semi-algebraic half-spaces $\{\theta = (\phi, \upsilon) : g_{\mathcal{B}'}\left(Adj(G(\phi)_{\bullet\mathcal{B}'})_{i,\bullet} q(\phi, \upsilon) > 0\}$ and $\{\theta = (\phi, \upsilon) : g_{\mathcal{B}'}\left(Adj(G(\phi)_{\bullet\mathcal{B}'}\right)_{j,\bullet} q(\phi, \upsilon) > 0\}$ with $B_{\epsilon(\theta)}(\theta)$ are disjoint. Recognize that points (i) and (ii) imply condition 1 and point (iii) implies condition 2.
($\Leftarrow$): Since we know that $\mathcal{IR}_{\mathcal{B}'}$ is adjacent to another invariancy region, from the definition of adjacency we know $dim(\mathcal{IR}_{\mathcal{B}'}) \geq k-1$. Recognize that condition 2 of the proposition implies that for every $\theta \in \mathcal{IR}_{\mathcal{B}'}$, there exist $j \in \mathcal{B}' \setminus \{i\}$ and $\epsilon(\theta) > 0$ such that the open semi-algebraic half-spaces $\{\theta = (\phi, \upsilon) : g_{\mathcal{B}'}\left(Adj(G(\phi)_{\bullet\mathcal{B}'}\right)_{i,\bullet} q(\phi, \upsilon) > 0\}$ and $\{\theta = (\phi, \upsilon) : g_{\mathcal{B}'}\left(Adj(G(\phi)_{\bullet\mathcal{B}'}\right)_{j,\bullet} q(\phi, \upsilon) > 0\}$ are disjoint within $B_{\epsilon(\theta)}(\theta)$. This is enough to show that $dim(\mathcal{IR}_{\mathcal{B}'}) \leq k-1$. Hence, we have $dim(\mathcal{IR}_{\mathcal{B}'}) = k-1$. □

The work done in the proof of Proposition 4.9 does provide some insight into which indices in a basis whose invariancy region is $(k-1)$-dimensional ought to be pivoted on, but in order to effectively use the result of the proposition, we require several concepts from algebra. For this, we quote several definitions and theorems from Bôcher and Duval [7]. Many of the results we cite here are properties of polynomials of multiple variables. We note that although many of the cited results are given specifically for polynomials of three variables, as stated in [7], the concepts apply directly to polynomials in any number of variables, and we therefore present the more general version of these results. We also note that the author of [7] uses the term *vanishes* to indicate the property of equalling zero.

Definition 4.3 (Definition 2 of Section 60 of Bôcher and Duval [7]) A polynomial is said to be *reducible* if it is identically equal to the product of two polynomials, neither of which is a constant.

Recognize that Definition 4.3 implies that a polynomial is *irreducible* if it cannot be written as the product of two nonconstant polynomials.

Proposition 4.10 (Shortened version of Theorem 6 of Section 76 of Bôcher and Duval [7]) *A polynomial in k variables that is not identically zero can be resolved into the product of irreducible factors, none of which is constant.*

Proposition 4.11 (Theorem 8 of Section 76 of Bôcher and Duval [7]) *If p and q are two polynomials in k variables, which both vanish at the point* $(x_1^0, \ldots, x_k^0)$ *and of which q is irreducible, and if in the neighborhood N of* $(x_1^0, \ldots, x_k^0)$ *p vanishes at all points at which q vanishes, then q is a factor of p.*

Definition 4.4 (Bôcher and Duval [7]) By the *greatest common divisor* of two polynomials is meant their common factor of greatest degree.

We now make the following important observations based on these definitions and propositions.

Observation 4.5 *The converse of Proposition* 4.11 *holds if we remove the condition that the polynomial q be irreducible.*

Observation 4.6 *Suppose the polynomial q in Proposition* 4.11 *was not assumed to be irreducible, then the claim "q is a factor of p" can be replaced with "the greatest common divisor of q and p has a nonconstant factor that vanishes at* $(x_1^0, \ldots, x_k^0)$*."*

Based on the results of Observation 4.5 and 4.6, we introduce some additional notation. Given a f.c.b. $\mathcal{B}$ and distinct $i, j \in \mathcal{B}$, we define

$$GCD(\mathcal{B}, i, j) :=$$

$$\text{Greatest common divisor of } (Adj(G(\phi)_{\bullet\mathcal{B}})_{i,\bullet}\, q(\phi, \upsilon) \text{ and } (Adj(G(\phi)_{\bullet\mathcal{B}})_{j,\bullet}\, q(\phi, \upsilon). \tag{4.20}$$

Then for each f.c.b. $\mathcal{B}$ and each index $i \in \mathcal{B}$, we define the set $\mathrm{D}^i_{\mathcal{B}}$ so that

$$\mathrm{D}^i_{\mathcal{B}} := \emptyset \text{ if } dim(\mathcal{IR}_{\mathcal{B}}) = k, \tag{4.21}$$

and

$$\mathrm{D}^i_{\mathcal{B}} := \big\{ j \in \mathcal{B} : GCD(\mathcal{B}, i, j) \text{ is a nonconstant polynomial and}$$
$$(\mathcal{IR}_{\mathcal{B}} \cap \hbar^i_{\mathcal{B}}) \subseteq (\mathcal{IR}_{\mathcal{B}} \cap GCD(\mathcal{B}, i, j))\big\} \text{ otherwise.} \tag{4.22}$$

In order to clarify these notations and highlight their importance in solving mpLCP, we provide the following small example. Suppose we are given an instance of mpLCP in which $\Theta = [-4, 4]^2$, and in the process of partitioning Θ, we discover a f.c.b. $\mathcal{B}$ with invariancy region

$$\mathcal{IR}_{\mathcal{B}} = \left\{ \theta \in \Theta : \begin{array}{l} w_1 = (\theta_1 - 2)(\theta_2 + 3) \geq 0 \\ w_2 = (\theta_1 - 2)(\theta_2 - 3) \geq 0 \\ w_3 = -\theta_1 - \theta_2 + 4 \geq 0 \\ w_4 = \theta_1 + \theta_2 \geq 0 \end{array} \right\}.$$

This invariancy region is displayed in Figure 4.3.

Recognize from Figure 4.3 that $dim(\mathcal{IR}_{\mathcal{B}}) = k - 1$ even though there is no defining constraint of $\mathcal{IR}_{\mathcal{B}}$ for which the LHS is a constant multiple of another constraint's LHS. Examples such as this one motivate our use of the sets $GCD(\mathcal{B}, i, j)$ and $\mathrm{D}^i_{\mathcal{B}}$. Notice that in this example we have $GCD(\mathcal{B}, w_1, w_2) = \theta_1 - 2$, while the greatest common divisor for each other pair of variables is a constant. Further recognize that $\hbar^{w_1}_{\mathcal{B}} \cap \mathcal{IR}_{\mathcal{B}} = \hbar^{w_2}_{\mathcal{B}} \cap \mathcal{IR}_{\mathcal{B}}$. Hence, we can see that $\mathrm{D}^{w_1}_{\mathcal{B}} = \{w_2\}$, $\mathrm{D}^{w_2}_{\mathcal{B}} = \{w_1\}$, $\mathrm{D}^{w_3}_{\mathcal{B}} = \emptyset$, and $\mathrm{D}^{w_4}_{\mathcal{B}} = \emptyset$. This small example provides some insight into the types of situations in which $(k-1)$-dimensional regions can arise. In order to be able to study these situations further, we need to

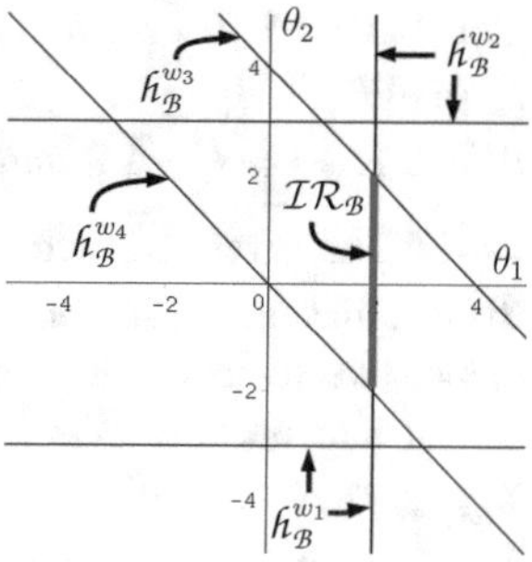

Fig. 4.3 Example of a $(k-1)$-dimensional region

develop a method for constructing the set $\mathrm{D}^i_{\mathcal{B}}$ for a given basis $\mathcal{B}$ and index $i \in \mathcal{B}$. One strategy for doing this is to first compute $GCD(\mathcal{B}, i, j)$ for each $j \in \mathcal{B} \setminus \{i\}$, and then if $GCD(\mathcal{B}, i, j)$ is a nonconstant polynomial, check to see if there is a point in $\mathcal{IR}_{\mathcal{B}}$ that lies on the surface $(Adj(G(\phi)_{\bullet\mathcal{B}}))_{i\bullet}q(\phi, \upsilon) = 0$ but at which $GCD(\mathcal{B}, i, j)$ is nonzero. If no such point exists, include j in $\mathrm{D}^i_{\mathcal{B}}$. Otherwise, do not. Recognize that one way to determine whether such a point exists is to determine the maximum value of $GCD(\mathcal{B}, i, j)^2$ for points in $\mathcal{IR}_{\mathcal{B}}$ that lie on the surface $(Adj(G(\phi)_{\bullet\mathcal{B}}))_{i\bullet}q(\phi, \upsilon) = 0$. If this maximum value is zero, include j in $\mathrm{D}^i_{\mathcal{B}}$. Otherwise, do not. This strategy motivates the following proposition.

Proposition 4.12 *Given a f.c.b. $\mathcal{B}$ such that $dim(\mathcal{IR}_{\mathcal{B}}) = k - 1$ and an index $i \in \mathcal{B}$, we have that $j \in \mathcal{B}\setminus\{i\}$ is in $\mathrm{D}^i_{\mathcal{B}}$ if and only if $GCD(\mathcal{B}, i, j)$ is a nonconstant polynomial and the following NLP has an optimal value of zero:*

$$
\begin{aligned}
NLP_G(\mathcal{B}, i, j) := \max_{\lambda, \phi, \upsilon} \quad & \lambda \\
\text{s.t.} \quad & g_{\mathcal{B}}\,(Adj(G(\phi)_{\bullet\mathcal{B}}))_{\ell\bullet}q(\phi, \upsilon) \geq 0 \;\; \forall \ell \in \left(\mathcal{B} \setminus (Z_{\mathcal{B}} \cup \{i\})\right) \\
& (Adj(G(\phi)_{\bullet\mathcal{B}}))_{i\bullet}q(\phi, \upsilon) = 0 \\
& GCD(\mathcal{B}, i, j)^2 \geq \lambda \\
& \phi \in \Phi, \;\; \upsilon \in \Upsilon
\end{aligned}
\tag{4.23}
$$

Proof It is clear from (4.21) and (4.22) that j cannot be in $\mathrm{D}^i_{\mathcal{B}}$ unless $GCD(\mathcal{B}, i, j)$ is a nonconstant polynomial. Recognize the similarity between the second condition given in (4.22) and (4.6). Hence, the remainder of the proof is analogous to the proof of Proposition 4.5. □

The algebraic properties of polynomials that we have now introduced, together with the subsequent observations we have made and the result of Proposition 4.12, lead to the following corollary that plays an important role in handling $(k-1)$-dimensional invariancy regions.

Corollary 4.5 *Let distinct f.c.b.'s $\mathcal{B}'$ and $\mathcal{B}$ be given that satisfy: (i) $dim(\mathcal{IR}_{\mathcal{B}}) = k$, (ii) $|\mathcal{B}' \cap \mathcal{B}| \geq h-2$, and (iii) $\mathcal{IR}_{\mathcal{B}}$ is adjacent to $\mathcal{IR}_{\mathcal{B}'}$ along $\hbar^i_{\mathcal{B}'}$. Then $dim\left(\mathcal{IR}_{\mathcal{B}'}\right) = k-1$ if and only if $D^i_{\mathcal{B}'} \neq \emptyset$.*

Proof The reverse direction of the proof follows directly from (4.21), (4.22), and the fact that since $\mathcal{IR}_{\mathcal{B}}$ is adjacent to $\mathcal{IR}_{\mathcal{B}'}$ along $\hbar^i_{\mathcal{B}'}$ we know $dim(\mathcal{IR}_{\mathcal{B}'}) \geq k-1$. Hence, we focus on the forward direction and assume that $dim\left(\mathcal{IR}_{\mathcal{B}'}\right) = k-1$. Recall that because $dim\left(\mathcal{IR}_{\mathcal{B}'}\right) = k-1$ and $\mathcal{IR}_{\mathcal{B}}$ is adjacent to $\mathcal{IR}_{\mathcal{B}'}$ along $\hbar^i_{\mathcal{B}'}$, we have that for all $\theta = (\phi, \upsilon) \in \mathcal{IR}_{\mathcal{B}'}$, $\left(Adj(G(\phi')_{\bullet\mathcal{B}'})\right)_{i,\bullet} q(\phi', \upsilon') = 0$. Thus, using the vocabulary of [7], condition (1) of Proposition 4.9 implies that there exist $j \in \mathcal{B}' \setminus \{i\}$ and $\theta' \in \mathcal{IR}_{\mathcal{B}'}$ such that in a neighborhood around θ', $\left(Adj(G(\phi')_{\bullet\mathcal{B}'})\right)_{j,\bullet} q(\phi', \upsilon')$ vanishes whenever $\left(Adj(G(\phi')_{\bullet\mathcal{B}'})\right)_{i,\bullet} q(\phi', \upsilon')$ vanishes. From Proposition 4.11 and Observation 4.6 and using (4.20), we conclude that: (i) $GCD(\mathcal{B}', i, j)$ is a nonconstant polynomial and (ii) $(\mathcal{IR}_{\mathcal{B}} \cap \hbar^i_{\mathcal{B}}) \subseteq (\mathcal{IR}_{\mathcal{B}} \cap GCD(\mathcal{B}, i, j))$. Hence, we observe from (4.22) that $D^i_{\mathcal{B}'}$ will not be empty. □

The result of Corollary 4.5 now shows us that, given a f.c.b. $\mathcal{B}$ for which $\mathcal{IR}_{\mathcal{B}}$ is $(k-1)$-dimensional and an $i \in \mathcal{B}$ for which $dim\left(\mathcal{IR}_{\mathcal{B}} \cap \hbar^i_{\mathcal{B}}\right) = k-1$, only pivots involving indices in $D^i_{\mathcal{B}} \subseteq \left(\mathcal{B} \setminus \{i\}\right)$ may yield new bases whose associated invariancy regions are adjacent to $\mathcal{IR}_{\mathcal{B}}$ and at least $(k-1)$-dimensional. One of the final theoretical results we need before being able to present an outline of the procedure FINDADJACENTKMINUS1 is a strategy for determining whether or not a pivot on an index in $D^i_{\mathcal{B}}$ will yield an adjacent invariancy region. For this purpose, we introduce the following proposition and its corollary.

Proposition 4.13 *Let distinct f.c.b.'s $\mathcal{B}$ and $\mathcal{B}'$ be given for which $dim(\mathcal{IR}_{\mathcal{B}}) = k-1$ and $|\mathcal{B} \cap \mathcal{B}'| \geq h-2$. Furthermore, let $\ell \in \mathcal{B}$ be an index for which $dim\left(\mathcal{IR}_{\mathcal{B}} \cap \hbar^\ell_{\mathcal{B}}\right) = k-1$. Then $\mathcal{IR}_{\mathcal{B}'}$ is adjacent to $\mathcal{IR}_{\mathcal{B}}$ along $\hbar^\ell_{\mathcal{B}}$ if and only if there exists $i \in D^\ell_{\mathcal{B}}$ such that one of the following conditions holds:*

1. *$\mathcal{B}' = (\mathcal{B} \setminus \{i\}) \cup \{\bar{\imath}\}$ and $(T_{\mathcal{B}}(\phi))_{i,\bar{\imath}} \not\equiv 0$.*
2. *There exists an additional index $j \in \mathcal{B} \setminus \{i\}$ such that $\mathcal{B}' = (\mathcal{B} \setminus \{i, j\}) \cup \{\bar{\imath}, \bar{j}\}$, $(T_{\mathcal{B}}(\phi))_{i,\bar{j}} \equiv 0$, and the following NLP has a strictly positive optimal value:*
$NLP_{A2}(\mathcal{B}, \bar{\imath}, \bar{j}) :=$

$$\begin{array}{lll} \max\limits_{\lambda,\phi,\upsilon} & \lambda & \\ s.t. & g_{\mathcal{B}}\,(Adj(G(\phi)_{\bullet\mathcal{B}}))_{j\bullet}\,G(\phi)_{\bullet\bar{\imath}} \geq \lambda & \\ & g_{\mathcal{B}}\,(Adj(G(\phi)_{\bullet\mathcal{B}}))_{\xi\bullet}\,q(\phi,\upsilon) \geq \lambda & \forall \xi \in \left(\mathcal{B} \setminus \left(Z_{\mathcal{B}} \cup H^{i}_{\mathcal{B}} \cup \{i\}\right)\right) \\ & (Adj(G(\phi)_{\bullet\mathcal{B}}))_{i\bullet}\,q(\phi,\upsilon) = 0 & \\ & g_{\mathcal{B}'}\left(Adj(G(\phi)_{\bullet\mathcal{B}'})\right)_{\xi\bullet}\,q(\phi,\upsilon) \geq \lambda & \forall \xi \in \left(\mathcal{B}' \setminus \left(Z_{\mathcal{B}'} \cup H^{\bar{\jmath}}_{\mathcal{B}'} \cup \{\bar{\jmath}\}\right)\right) \\ & \phi \in \Phi,\ \upsilon \in \Upsilon & \end{array} \tag{4.24}$$

Proof It is clear from Observation 4.6, (4.20), (4.21), and (4.22) that $dim(\mathcal{IR}_{\mathcal{B}} \cap \hbar^{\ell}_{\mathcal{B}} \cap \hbar^{i}_{\mathcal{B}}) = k-1$ if and only if $i \in \mathrm{D}^{\ell}_{\mathcal{B}}$. The remainder of the proof is analogous to that of Proposition 4.7. □

Corollary 4.6 *Let a f.c.b. $\mathcal{B}$ be given, together with distinct $i, j \in \mathcal{B}$. Consider $\mathcal{B}' = \left(\mathcal{B} \setminus \{i, j\}\right) \cup \{\bar{\imath}, \bar{\jmath}\}$. If $(T_{\mathcal{B}}(\phi))_{j,\bar{\imath}}$ is a nonpositive constant, then $NLP_{A2}(\mathcal{B}, i, j)$ cannot have a strictly positive optimal value.*

Proof The proof is analogous to that of Corollary 4.3. □

The result of Proposition 4.13 now provides us with a strategy so that, given a $(k-1)$-dimensional invariancy region $\mathcal{IR}_{\mathcal{B}}$ together with an $i \in \mathcal{B}$ for which we know $dim\left(\mathcal{IR}_{\mathcal{B}} \cap \hbar^{i}_{\mathcal{B}}\right) = k-1$, we can find the set of all previously undiscovered f.c.b.'s whose associated invariancy regions are adjacent to $\mathcal{IR}_{\mathcal{B}}$ along $\hbar^{i}_{\mathcal{B}}$ and have dimension at least $k-1$. We present this strategy in Algorithm 6.

Recognize the similarity of Algorithms 4 and 6. In fact, from line 6 on, Algorithm 6 mirrors Algorithm 4 almost exactly. The only significant difference between the two algorithms is that in Algorithm 6, prior to searching for adjacent regions, we must first construct the set of indices that are candidates for pivoting. This is done on lines 2–4.

Also recognize that, as constructed, Algorithm 6 is recursive. The reason for this is twofold: (i) the algorithm searches for f.c.b.'s whose associated invariancy regions are adjacent to a given $(k-1)$-dimensional region and (ii) there is a possibility that any newly discovered region might also be $(k-1)$-dimensional. We do note, however, that neither Algorithm 4 nor Algorithm 6 (the two routines that call FINDADJACENTKMINUS1) allows for the consideration of any f.c.b. more than once, and as a result, Algorithm 6 will always execute in finite time because there are a finite number of f.c.b.'s.

There is one last result that we need to establish before completing this section. Given a f.c.b. $\mathcal{B}$ whose invariancy region is $(k-1)$-dimensional, we need to establish that the index $\ell \in \mathcal{B}$ that we will use in a call to FINDADJACENTKMINUS1$(\mathcal{B}, \ell, \mathscr{B})$ satisfies the property that $dim\left(\mathcal{IR}_{\mathcal{B}} \cap \hbar^{\ell}_{\mathcal{B}}\right) = k-1$. Without this property, we cannot guarantee that Algorithm 6 will execute as intended. The following proposition contains the necessary result.

Algorithm 6 FINDADJACENTKMINUS1($\mathcal{B}$,ℓ,$\mathscr{B}$)—Find all previously undiscovered invariancy regions that are adjacent to a given $(k-1)$-dimensional invariancy region

Input: A f.c.b. $\mathcal{B}$ such that $dim(\mathcal{IR}_{\mathcal{B}}) = k - 1$, an index $\ell \in \mathcal{B}$ for which $dim\left(\mathcal{IR}_{\mathcal{B}} \cap \hbar^{\ell}_{\mathcal{B}}\right) = k - 1$ and the set $\mathscr{B}$ of previously discovered bases.
Output: The set $\mathcal{S}'$ of f.c.b.'s with associated invariancy regions that are adjacent to $\mathcal{IR}_{\mathcal{B}}$ and an updated version of $\mathscr{B}$.

1: Set $\mathcal{S}' = \emptyset$ and $\mathrm{D}^{\ell}_{\mathcal{B}} = \emptyset$.
2: **for** $j \in \mathcal{B} \setminus \{\ell\}$ **do** compute $GCD(\mathcal{B}, \ell, j)$.
3: **if** $GCD(\mathcal{B}, \ell, j)$ is a nonconstant polynomial **then** solve $NLP_G(\mathcal{B}, \ell, j)$ to obtain an optimal solution $(\lambda^*, \phi^*, \upsilon^*)$.
4: **if** $\lambda^* = 0$ **then** add j to $\mathrm{D}^{\ell}_{\mathcal{B}}$.
5: **for** $i \in \mathrm{D}^{\ell}_{\mathcal{B}}$ **do** set $\mathcal{B}' = (\mathcal{B} \setminus \{i\}) \cup \{\bar{\imath}\}$.
6: **if** $\mathcal{B}' \notin \mathscr{B}$ and $(T_{\mathcal{B}}(\phi))_{i,\bar{\imath}} \not\equiv 0$ **then** set $\mathscr{B} = \mathscr{B} \cup \{\mathcal{B}'\}$ and let $d =$ GETINVARIANCYREGIONDIMENSION($\mathcal{B}'$).
7: **if** $d = k$ **then** set $\mathcal{S}' = \mathcal{S}' \cup \{\mathcal{B}'\}$.
8: **else** set $(\mathcal{S}'', \mathscr{B}) =$ FINDADJACENTKMINUS1($\mathcal{B}', \bar{\imath}, \mathscr{B}$) and $\mathcal{S}' = \mathcal{S}' \cup \mathcal{S}''$.
9: **else**
10: **for** $j \in \mathcal{B} \setminus \{i\}$ **do** set $\mathcal{B}'' = (\mathcal{B} \setminus \{i, j\}) \cup \{\bar{\imath}, \bar{\jmath}\}$.
11: **if** $\mathcal{B}'' \notin \mathscr{B}$ and $(T_{\mathcal{B}}(\phi))_{j,\bar{\imath}}$ either has degree of at least 1 or is a strictly positive constant **then** compute $T_{\mathcal{B}''}(\phi)$ and solve $NLP_{A2}(\mathcal{B}, i, j)$ to obtain an optimal solution $(\lambda^*, \phi^*, \upsilon^*)$.
12: **if** $\lambda^* > 0$ **then** set $\mathscr{B} = \mathscr{B} \cup \{\mathcal{B}''\}$ and let $d =$ GETINVARIANCYREGIONDIMENSION($\mathcal{B}''$).
13: **if** $d = k$ **then** set $\mathcal{S}' = \mathcal{S}' \cup \{\mathcal{B}''\}$.
14: **else** set $(\mathcal{S}'', \mathscr{B}) =$ FINDADJACENTKMINUS1($\mathcal{B}'', \bar{\jmath}, \mathscr{B}$) and $\mathcal{S}' = \mathcal{S}' \cup \mathcal{S}''$.
15: Return $\mathcal{S}'$ and $\mathscr{B}$.

Proposition 4.14 *Let distinct f.c.b.'s $\mathcal{B}$ and $\mathcal{B}'$ be given such that: (i) $\mathcal{B}$ has been obtained via a pivot from $\mathcal{B}'$, (ii) $dim(\mathcal{IR}_{\mathcal{B}}) = k-1$, (iii) $|\mathcal{B} \cap \mathcal{B}'| \geq h-2$, and (iv) $\mathcal{IR}_{\mathcal{B}}$ and $\mathcal{IR}_{\mathcal{B}'}$ are adjacent across $\hbar^{i}_{\mathcal{B}'}$ for some $i \in \mathcal{B}'$. The following hold:*

1. *If $\mathcal{B} = (\mathcal{B}' \setminus \{i\}) \cup \{\bar{\imath}\}$, then $dim\left(\mathcal{IR}_{\mathcal{B}} \cap \hbar^{\bar{\imath}}_{\mathcal{B}}\right) = k - 1$.*
2. *If $\mathcal{B} = (\mathcal{B}' \setminus \{i, j\}) \cup \{\bar{\imath}, \bar{\jmath}\}$ for some $j \in \mathcal{B}'$, then $dim\left(\mathcal{IR}_{\mathcal{B}} \cap \hbar^{\bar{\jmath}}_{\mathcal{B}}\right) = k - 1$.*

Proof First assume that $\mathcal{B} = (\mathcal{B}' \setminus \{i\}) \cup \{\bar{\imath}\}$. Hence, $\mathcal{B}$ was obtained from $\mathcal{B}'$ using a diagonal pivot as outlined in condition (1) of either Proposition 4.7 or 4.13. Recognize that by reversing the roles of $\mathcal{B}$ and $\mathcal{B}'$ in Proposition 4.13, it is clear that in this case $\mathcal{IR}_B$ and $\mathcal{IR}_{B'}$ are adjacent along $\hbar^{i}_{\mathcal{B}'}$ if and only if $\mathcal{IR}_B$ and $\mathcal{IR}_{B'}$ are adjacent along $\hbar^{\bar{\imath}}_{\mathcal{B}}$. The claim of the proposition follows.

Now assume that there exists $j \in \mathcal{B}'$ such that $\mathcal{B} = (\mathcal{B}' \setminus \{i, j\}) \cup \{\bar{\imath}, \bar{\jmath}\}$. Then $\mathcal{B}$ was obtained from $\mathcal{B}'$ using an exchange pivot as outlined in condition (2) of either Proposition 4.7 or 4.13. In this case, analogous arguments to those used in the proof of Proposition 4.7 can be used to show that $dim\left(\mathcal{IR}_{\mathcal{B}} \cap \hbar^{i}_{\mathcal{B}'} \cap \hbar^{\bar{\jmath}}_{\mathcal{B}}\right) = k-1$. Thus, the claim of the proposition holds. □

We now return our attention to Example 2.2. Previously we saw that the optimal value of $NLP_D(\mathcal{B}_2^{2.2})$ was zero and thus $dim(\mathcal{IR}_{\mathcal{B}_2^{2.2}}) = k - 1$. Our previous work and Proposition 4.14 also show that $dim\left(\mathcal{IR}_{\mathcal{B}_2^{2.2}} \cap \hbar^{z_2}_{\mathcal{B}_2^{2.2}}\right) = k - 1$. Thus, continuing the execution of the procedure GETADJACENTREGIONSACROSS($\mathcal{B}_0^{2.2}$), we now run FINDADJACENTKMINUS1($\mathcal{B}_2^{2.2}$,z_2,$\mathscr{B}$). In doing so we find that $GCD(B_2^{2.2}, z_2, w_1) = \frac{1}{15}$, $GCD(B_2^{2.2}, z_2, z_3) = \frac{1}{15}$, and $GCD(B_2^{2.2}, z_2, w_4) = 3\upsilon + \frac{17}{3}$. As $GCD(B_2^{2.2}, z_2, w_4)$ is the only one of these that is a nonconstant polynomial, we now solve $NLP_G(B_2^{2.2}, z_2, w_4)$. This gives an approximate optimal solution of $(\lambda^*, \phi^*, \upsilon^*) = (0, -0.503, -1.8889)$, which shows that $w_4 \in \mathrm{D}^{z_2}_{\mathcal{B}_2^{2.2}}$. We observe from $T_{\mathcal{B}_2^{2.2}}(\phi)$, found in Table B.3 in Appendix B, that the only possible pivot from $\mathcal{B}_2^{2.2}$ that involves w_4 is the exchange pivot that also involves w_1. This pivot results in obtaining $\mathcal{B}_3^{2.2} = \{z_1, z_2, z_3, z_4\}$. In order to ensure that $\mathcal{B}_3^{2.2}$ yields an invariancy region adjacent to $\mathcal{IR}_{\mathcal{B}_2^{2.2}}$, we solve $NLP_{A2}(\mathcal{B}_2^{2.2}, w_4, w_1)$, which has an approximate optimal solution of $(\lambda^{**}, \phi^{**}, \upsilon^{**}) = (1.4815, 0.5926, -1.8889)$. Since $\lambda^{**} > 0$, we conclude that $\mathcal{IR}_{\mathcal{B}_3^{2.2}}$ is, in fact, adjacent to $\mathcal{IR}_{\mathcal{B}_2^{2.2}}$ and we proceed by calling GETINVARIANCYREGIONDIMENSION($\mathcal{B}_3^{2.2}$). Table B.3 can be used to construct $NLP_D(\mathcal{B}_3^{2.2})$, which has an approximate optimal solution of $(\lambda^{***}, \phi^{***}, \upsilon^{***}) = (3.1434, -0.7623, -1.5396)$. Thus, since $\lambda^{***} > 0$, we know $dim\left(\mathcal{IR}_{\mathcal{B}_3^{2.2}}\right) = k$. This now causes the execution of GETINVARIANCYREGIONDIMENSION($\mathcal{B}_3^{2.2}$) to terminate. Hence, we return to the execution of GETADJACENTREGIONSACROSS($\mathcal{B}_0^{2.2}$). Note, however, that because $dim\left(\mathcal{IR}_{\mathcal{B}_3^{2.2}}\right) = k$, the current iteration of the for loop on line 7 of Algorithm 4 now ceases. We therefore begin the next (and final) iteration of this loop by considering w_4. The if statement on line 9 reveals that the only exchange pivot we must consider involves w_4 and w_1. An optimal solution to $NLP_A(\mathcal{B}_0^{2.2}, w_4, w_1)$ is approximately $(-0.3611, -1.25, -1.8889)$, which shows that basis $\{w_1, z_2, w_3, z_4\}$ does not yield an invariancy region adjacent to $\mathcal{IR}_{\mathcal{B}_0^{2.2}}$ across $\hbar^{w_4}_{\mathcal{B}_0^{2.2}}$. We have now completed the execution of GETADJACENTREGIONSACROSS($\mathcal{B}_0^{2.2}$). To complete our partition of Θ, we must now return to Algorithm 1 and process the next f.c.b. on the stack $\mathcal{S}$, which is $\mathcal{B}_3^{2.2}$.

We leave the remainder of the consideration of Example 2.2 to the reader, as the rest of the steps for partitioning Θ are analogous to steps we have already shown. Recall, though, that the solution to both examples can be observed in Tables 4.1 and 4.2. In the next chapter we discuss a technique for determining an initial f.c.b. with a full dimensional invariancy region.

Chapter 5
Phase 1: Determining an Initial Feasible Solution

The purpose of this chapter is to present a strategy for computing an initial feasible solution to mpLCP (1.1) with which we can begin the partitioning of the parameter space Θ. Thus, we aim to discover a f.c.b. $\mathcal{B}_0$ such that $dim\left(\mathcal{IR}_{\mathcal{B}_0}\right) = k$. Throughout this chapter, we work under the assumption that $\mathbf{0} \in \Theta$. This assumption is not restrictive, though, as it can be easily achieved by a straightforward translation. We now define the following augmented phase 1 multiparametric LCP, denoted as mpLCP_{ph1}:

$$\begin{aligned} w - M(\phi)z &= q(\phi, \upsilon) + r\rho \\ w^\top z &= 0 \\ w, z &\geq 0 \end{aligned} \tag{5.1}$$

Here, $\rho \in \mathbb{R}$ is an additional parameter and $r \in \mathbb{R}^h$ is defined so that, if we represent $q(\phi, \upsilon)$ as $q + \triangle Q_{\bullet U}\phi + \triangle Q_{\bullet V}$, we have

$$r_i = \begin{cases} |q_i| + 1 & \text{if } q_i \leq 0 \\ 0 & \text{otherwise} \end{cases} \quad \text{for each } i \in \{1, \dots, h\}. \tag{5.2}$$

Clearly mpLPC_{ph1} (5.1) is a variant of mpLCP (1.1) in which k is replaced by $k+1$. As a result, previously presented definitions and theory for mpLCP all directly apply to mpLPC_{ph1}. In the remainder of this chapter, we use the superscript $ph1$ to denote the phase 1 analogues of various sets and other notations we defined in Chapter 4. For example, the notations $\mathcal{IR}_{\mathcal{B}}^{ph1}$ and $(\hbar_{\mathcal{B}}^i)^{ph1}$ represent the respective analogues of $\mathcal{IR}_{\mathcal{B}}$ and $\hbar_{\mathcal{B}}^i$ for mpLPC_{ph1}.

We point out that we place no restrictions on the value of ρ. Consequently, an unbounded solution may exist for the phase 1 counterpart to one or more of the NLPs presented for phase 2. This is not problematic, however, since the optimal value of

N. Adelgren, *Advancing Parametric Optimization*, SpringerBriefs in Optimization,
https://doi.org/10.1007/978-3-030-61821-6_5

an unbounded maximization (minimization) problem is strictly positive (negative), which is the condition that we seek for most of the presented NLPs.

Proposition 5.1 *The complementary basis*

$$\mathcal{B}^* := \{1, \ldots, h\} \tag{5.3}$$

is feasible to $mpLCP_{ph1}$ (5.1)*, and* $\mathcal{IR}_{\mathcal{B}^*}^{ph1}$ *is full dimensional.*

Proof Since $G(\phi)_{\bullet\mathcal{B}^*} = I$, we have

$$\mathcal{IR}_{\mathcal{B}^*}^{ph1} = \{(\phi, \upsilon, \rho) \in \Phi \times \Upsilon \times \mathbb{R} : q(\phi, \upsilon) + r\rho \geq 0\}.$$

From this and (5.2), we also have $(\phi, \upsilon, \rho) = (\mathbf{0}, \mathbf{0}, 1) \in \mathcal{IR}_{\mathcal{B}^*}^{ph1}$. Hence, $\mathcal{IR}_{\mathcal{B}^*}^{ph1} \neq \emptyset$, which further shows that $\mathcal{B}^*$ is feasible to mpLCP$_{ph1}$. Moreover, the system $q(\phi, \upsilon) + r\rho \geq 0$ is satisfied strictly at $(\phi, \upsilon, \rho) = (\mathbf{0}, \mathbf{0}, 1)$, and so there must exist $\epsilon > 0$ such that $B_\epsilon((\mathbf{0}, \mathbf{0}, 1)) \subset \mathcal{IR}_{\mathcal{B}^*}^{ph1}$, i.e., $\mathcal{IR}_{\mathcal{B}^*}^{ph1}$ is full dimensional. □

The result of Proposition 5.1 is extremely useful as it shows the immediate availability of a full dimensional invariancy region for mpLCP$_{ph1}$. This result also provides the following simple strategy for computing an initial basis $\mathcal{B}_0$. Determine the k-dimensional boundaries of $\mathcal{IR}_{\mathcal{B}^*}^{ph1}$, compute the bases whose phase 1 invariancy regions are adjacent to $\mathcal{IR}_{\mathcal{B}^*}^{ph1}$ across each such boundary, and then repeat this procedure for each newly discovered invariancy region. For each newly discovered basis $\mathcal{B}$, solve $NLP_D(\mathcal{B})$ (4.19) to determine the dimension of $\mathcal{IR}_{\mathcal{B}}$. Continue partitioning $\Theta \times \mathbb{R}$ in the same fashion that we suggested for partitioning Θ in Chapter 4, and cease computation when a basis is found whose associated phase 2 invariancy region is full dimensional. If this procedure yields no such basis, then no such basis exists. Note that if no full dimensional invariancy region is found, it is unnecessary to search for $(k-1)$-dimensional invariancy regions. This procedure is clearly a brute-force method, and is therefore not very efficient in practice, but it does serve as a good foundation for the strategy we ultimately propose.

We note here the similarity between the strategy described above and that of partitioning Θ as described in Chapter 4. As these procedures are very similar, they share similar shortcomings. Namely, just as the presented strategy for partitioning Θ relies on Assumption 1.2, the strategies presented in this chapter rely on the following assumption.

Assumption 5.1 *We assume that there exists a set*

$$\Gamma \subseteq \{(\theta, \rho) \in \Theta \times \mathbb{R} : \ (\theta, \rho) \textit{ is feasible to } mpLCP_{ph1}\}$$

such that $(\mathbf{0}, 1) \in \Gamma$*,* $B_\epsilon(\theta') \subseteq \Gamma$ *for some* $\theta' \in \Theta$ *and* $\epsilon > 0$*, and* Γ *is connected in its interior.*

Under Assumption 5.1, notice that the existence of such a set Γ ensures that it is possible to use strategies similar to those described in Chapter 4 in order to partition Γ into a set of $(k+1)$-dimensional sets and, moreover, at least one of those sets must have k-dimensional intersection with the hyperplane $\rho = 0$. Thus, under Assumption 5.1, it must be possible to find a f.c.b. with which we can begin the phase 2 algorithm presented in Chapter 4.

We now continue our previous discussion and work to develop a more efficient strategy for phase 1 that does not require generating a full partition of $\Theta \times \mathbb{R}$. Recognize that for each phase 1 invariancy region $\mathcal{IR}_{\mathcal{B}}^{ph1}$, the phase 2 invariancy region $\mathcal{IR}_{\mathcal{B}}$ is the intersection of $\mathcal{IR}_{\mathcal{B}}^{ph1}$ with the hyperplane $\rho = 0$. Hence, if we can determine that the likelihood of discovering adjacent invariancy regions whose intersections with the hyperplane $\rho = 0$ have dimension k is higher for some k-dimensional boundaries of an invariancy region than for others, we can improve upon the technique discussed above. With this in mind, consider the following NLP and the subsequent related proposition:

$$
\begin{aligned}
NLP_S(\mathcal{B}) := \min_{\phi,\upsilon,\rho} \quad & \rho \\
\text{s.t.} \quad & g_{\mathcal{B}} Adj(G(\phi)_{\bullet\mathcal{B}})\,(q(\phi,\upsilon) + r\rho) \geq 0 \\
& \phi \in \Phi, \upsilon \in \Upsilon
\end{aligned}
\tag{5.4}
$$

Proposition 5.2 *If $M(\phi)$ is a Q_0 matrix for all $\phi \in \Phi$, then mpLCP (1.1) is feasible if and only if there exists a f.c.b. $\mathcal{B}$ for which $NLP_S(\mathcal{B})$ (5.4) has a nonpositive optimal value.*

Proof ($\Rightarrow$): If mpLCP (1.1) is feasible, then there exists a f.c.b. $\mathcal{B}'$ and $\hat{\theta} = (\hat{\phi}, \hat{\upsilon}) \in \Theta$ for which $g_{\mathcal{B}'} Adj(G(\hat{\phi})_{\bullet\mathcal{B}'})q(\hat{\phi}, \hat{\upsilon}) \geq 0$. Then $(\phi, \upsilon, \rho) = (\hat{\phi}, \hat{\upsilon}, 0)$ is feasible to $NLP_S(\mathcal{B}')$. The optimal value must therefore be nonpositive.
($\Leftarrow$): Recall from the proof of Proposition 5.1 that $NLP_S(\mathcal{B}^*)$ has a feasible solution in which $\rho = 1$. It is clear from this and the fact that mpLCP (1.1) is equivalent to mpLCP^{ph1} with ρ fixed to 0 that if there exists a basis $\mathcal{B}'$ for which $NLP_S(\mathcal{B}')$ is feasible for some $\hat{\rho} \leq 0$ then, since $\mathcal{K}(M(\phi))$ is convex for all ϕ such that $M(\phi)$ is Q_0, there must exist a basis $\mathcal{B}''$ for which $NLP_S(\mathcal{B}'')$ is feasible at $\rho = 0$. Hence, mpLCP is feasible. □

Given a f.c.b. $\mathcal{B}$, denote an optimal solution of $NLP_S(\mathcal{B})$ as $(\phi_{\mathcal{B}}^*, \upsilon_{\mathcal{B}}^*, \rho_{\mathcal{B}}^*)$ and define

$$
EQ_{\mathcal{B}} := \left\{ i \in \mathcal{B} : \left(Adj(G(\phi_{\mathcal{B}}^*)_{\bullet\mathcal{B}})\right)_{i,\bullet} \left(q(\phi_{\mathcal{B}}^*, \upsilon_{\mathcal{B}}^*) + r\rho_{\mathcal{B}}^*\right) = 0 \right\}, \tag{5.5}
$$

the set of indices in $\mathcal{B}$ whose corresponding defining constraints of $\mathcal{IR}_{\mathcal{B}}^{ph1}$ are binding at $(\phi_{\mathcal{B}}^*, \upsilon_{\mathcal{B}}^*, \rho_{\mathcal{B}}^*)$.

Proposition 5.3 *Assume that $M(\phi)$ is a Q_0 matrix for all $\phi \in \Phi$. Let a f.c.b. $\mathcal{B}$ be given, and let $(\phi_{\mathcal{B}}^*, \upsilon_{\mathcal{B}}^*, \rho_{\mathcal{B}}^*)$ be an optimal solution of $NLP_S(\mathcal{B})$ (5.4). If there is no*

$i \in EQ_{\mathcal{B}}$ for which a diagonal or exchange pivot can be made from $\mathcal{B}$ that involves index i, then the following hold:

- *If $\rho^*_{\mathcal{B}} > 0$, then mpLCP is infeasible.*
- *If $\rho^*_{\mathcal{B}} = 0$ and $dim\ (\mathcal{IR}_{\mathcal{B}}) < k$, then there is no f.c.b. $\mathcal{B}'$ such that $dim\left(\mathcal{IR}_{\mathcal{B}'}\right) = k$.*

Proof If for some $i \in \mathcal{B}$, there are no possible diagonal or exchange pivots that involve i, then the facet $cone\left(G(\phi)_{\bullet(\mathcal{B}\setminus\{i\})}\right)$ of the parametric complementary cone $\mathcal{C}_{\mathcal{B}}(\phi)$ is a boundary of $\mathcal{K}(M(\phi))$ for all $\phi \in \Phi$. Furthermore, because $\mathcal{K}(M(\phi))$ is convex for each ϕ such that $M(\phi)$ is a Q_0 matrix, all phase 1 invariancy regions lie in the same semi-algebraic half-space defined by the hypersurface $(\hbar^i_{\mathcal{B}})^{ph1}$ that contains $\mathcal{IR}^{ph1}_{\mathcal{B}}$. Therefore, since we have assumed that there is no $i \in EQ_{\mathcal{B}}$ for which a diagonal or exchange pivot can be made from $\mathcal{B}$ that involves index i, we have the following:

1. If the optimal value of $NLP_S(\mathcal{B})$ is strictly positive, no phase 1 invariancy region exists that intersects the hyperplane $\rho = 0$.
2. If the optimal value of $NLP_S(\mathcal{B})$ is zero, no phase 1 invariancy region other than $\mathcal{IR}^{ph1}_{\mathcal{B}}$ can have a nonempty intersection with the hyperplane $\rho = 0$.

Since $\mathcal{IR}_{\mathcal{B}}$ is the intersection of $\rho = 0$ and $\mathcal{IR}^{ph1}_{\mathcal{B}}$, the claim of the proposition follows. □

Proposition 5.3 provides two simplifications to the brute-force method. They are (i) the certification that only a subset of the k-dimensional boundaries of an invariancy region needs to be checked for adjacent invariancy regions and (ii) a stopping criterion under which one may conclude that either mpLCP (1.1) is infeasible or no full dimensional invariancy regions exist within Θ. Hence, our strategy for obtaining an initial invariancy region of full dimension is to follow the same procedure outlined in Chapter 4 for partitioning Θ, except, given any discovered f.c.b. $\mathcal{B}$, only consider pivoting on the indices in $\mathcal{B} \cap EQ_{\mathcal{B}}$. We then continue this modified partitioning procedure until we discover a f.c.b. $\mathcal{B}'$ for which either (i) $dim\left(\mathcal{IR}_{\mathcal{B}'}\right) = k$, thus ending phase 1 or (ii) there are no possible pivots from $\mathcal{B}'$ involving indices in $EQ_{\mathcal{B}'}$, in which case we conclude that no full dimensional invariancy regions exist.

The procedures described above are presented explicitly in Algorithm 7. For use within Algorithm 7 and the remainder of this chapter, we note that each of the procedures we outlined in an algorithm in Chapter 4 has a counterpart that is applicable for phase 1, with the exception of PARTITIONΘ, of course. For each procedure, we denote the phase 1 counterpart by appending "_PH1" to the name of the procedure. Many of these procedures, such as BUILDEANDH_PH1, are completely analogous to their phase 2 counterparts. However, some require slight modifications, and so for these, we will present updated algorithms throughout this chapter. Additionally, note that all of the NLPs presented in Chapter 4 (with the exception of NLP_{A2}) also have analogous counterparts that are applicable for phase

Algorithm 7 FINDINITIALBASIS—Find an initial f.c.b. having a full dimensional phase 2 invariancy region

Input: An instance of mpLCP.
Output: An initial f.c.b. $\mathcal{B}_0$ such that $dim\left(\mathcal{IR}_{\mathcal{B}_0}\right) = k$.

1: Let $\mathcal{S} = \{\mathcal{B}^*\}$ (see (5.3)) and $\mathscr{B} = \emptyset$.
2: **while** $\mathcal{S} \neq \emptyset$ **do** select $\mathcal{B}$ from $\mathcal{S}$.
3: Solve $NLP_S(\mathcal{B})$ to obtain an optimal solution $(\rho', \phi', \upsilon')$. Construct $EQ_{\mathcal{B}}$.
4: **if** $\rho' < 0$ **then** let d = GETINVARIANCYREGIONDIMENSION($\mathcal{B}$).
5: **if** $d = k$ **then** Return $\mathcal{B}$.
6: **else** let F = BUILDF_PH1($\mathcal{B}$, $\mathcal{B}$).
7: **for** $i \in$ F **do**
8: Let $(\mathcal{S}', \mathcal{S}'', \mathscr{B})$ = GETADJACENTREGIONSACROSS_PH1($\mathcal{B}, i, \mathscr{B}$) and set $\mathcal{S} = \mathcal{S} \cup \mathcal{S}'$.
9: **else**
10: Let $\mathscr{S} = \emptyset$ and F = BUILDF_PH1($\mathcal{B}$, $EQ_{\mathcal{B}}$).
11: **for** $i \in$ F **do**
12: Let $(\mathcal{S}', \mathcal{S}'', \mathscr{B})$ = GETADJACENTREGIONSACROSS_PH1($\mathcal{B}, i, \mathscr{B}$).
13: Set $\mathscr{S} = \mathscr{S} \cup \mathcal{S}' \cup \mathcal{S}''$ and $\mathcal{S} = \mathcal{S} \cup \mathcal{S}'$.
14: **if** $\mathscr{S} = \emptyset$ **then** STOP. There is no $\mathcal{B}'$ such that $dim\left(\mathcal{IR}_{\mathcal{B}'}\right) = k$.
15: **if** $\mathcal{S} = \emptyset$ **then** STOP. There is no $\mathcal{B}'$ such that $dim\left(\mathcal{IR}_{\mathcal{B}'}\right) = k$.

1. For each NLP, we denote the phase 1 analogue by adding a superscript of "ph1." Note that the only modification needed to obtain these NLPs is that in each, $q(\phi, \upsilon)$ be replaced by $q(\phi, \upsilon) + r\rho$.

On line 3 of Algorithm 7, we solve NLP_S and construct the set $EQ_{\mathcal{B}}$. Lines 4–8 handle the case in which the optimal value of NLP_S is strictly negative, and lines 9–14 handle the case in which it is not. Note that unlike the work of Adelgren and Wiecek [2], we cannot immediately conclude that our initial f.c.b. has been found if NLP_S has a strictly negative optimal value. This is because the invariancy regions considered herein are semi-algebraic (due to the presence of parameters in M), whereas the invariancy regions considered in [2] are polyhedral. Moreover, an algebraic or semi-algebraic set of dimension n can contain a point p such that for a given $\epsilon > 0$, we have $dim(p \cap B_\epsilon(p)) < n$. Examples of such sets are the so-called umbrellas (see, for example, [8]) such as Whitney's umbrella [55]. The presence of such points implies that a phase 1 invariancy region of full dimension could have nonempty intersection with both the half-spaces $\rho > 0$ and $\rho < 0$ and yet have an intersection with the hyperplane $\rho = 0$ that is less than k-dimensional. For these reasons, when we find a f.c.b. for which the optimal value of NLP_S is strictly negative, we use the routine GETINVARIANCYREGIONDIMENSION, presented in Chapter 4, to determine the dimension of its phase 2 invariancy region. If the dimension is full, we have found our initial f.c.b. and return it. Otherwise, we use lines 6–8 to compute adjacent regions.

When the optimal value of NLP_S is not strictly negative, lines 10–14 of Algorithm 7 employ the strategies resulting from Proposition 5.3. Lines 10–13 are used to determine adjacent regions across only the necessary subset of the boundaries of the current invariancy region, and line 14 issues a certificate that no full dimensional invariancy region exists in the case in which there are no such adjacent regions.

The correctness of Algorithm 7, namely that if any f.c.b. with full dimensional phase 2 invariancy region exists, then such a f.c.b. will be returned by Algorithm 7, is argued at the end of this chapter, in Proposition 5.5.

Note from Algorithm 7 that the routines BUILDF_PH1 and GETADJACENTREGIONSACROSS_PH1 differ from their phase 2 counterparts. BUILDF_PH1 takes an additional input, while GETADJACENTREGIONSACROSS_PH1 returns an additional output. We now present Algorithms 8 and 9, which describe BUILDF_PH1 and GETADJACENTREGIONSACROSS_PH1, respectively.

Observe from Algorithm 8 that the only difference between BUILDF_PH1 and BUILDF (Algorithm 2) is that line 2 of Algorithm 8 utilizes a subset of a given f.c.b. in place of the f.c.b. itself, as is used in Algorithm 2.

By observing Algorithm 9, one can see that the routine GETADJACENTREGIONSACROSS_PH1 is analogous to its phase 2 counterpart (Algorithm 4) with one minor exception. In GETADJACENTREGIONSACROSS_PH1, we keep track of f.c.b.'s that are already discovered, and thus in $\mathscr{B}$, but might yield regions adjacent to the current invariancy region across the specified hypersurface. The reason for keeping track of such f.c.b.'s is to justify the stopping criterion of Proposition 5.3 utilized on line 14 of Algorithm 7.

Recognize from Algorithm 9 that GETADJACENTREGIONSACROSS_PH1 calls two subroutines, GETINVARIANCYREGIONDIMENSION_PH1 and FINDADJACENTK_PH1. GETINVARIANCYREGIONDIMENSION_PH1 is completely analogous to its phase 2 counterpart (Algorithm 5). FINDADJACENTK_PH1, however, which is the phase 1 counterpart to FINDADJACENTKMINUS1 (Algorithm 6), must be modified. In fact, notice that FINDADJACENTK_PH1 takes two more inputs than FINDADJACENTKMINUS1. To see why this modification is necessary, consider the following situation. Suppose we have a f.c.b. $\mathcal{B}$ such that $\mathcal{IR}^{ph1}_{\mathcal{B}}$ is $(k+1)$-dimensional but is contained in the half-space $\rho \geq 0$ and $dim(\mathcal{IR}_{\mathcal{B}}) < k$. Further, suppose that we pivot on an index in $EQ_{\mathcal{B}}$ and obtain a new basis $\mathcal{B}'$ for which $dim(\mathcal{IR}^{ph1}_{\mathcal{B}'}) = k$ and the optimal value of $NLP_S(\mathcal{B}')$ is strictly negative. It is then possible to obtain a third basis $\mathcal{B}''$ by a pivot on an index in $EQ_{\mathcal{B}'}$ such

---

Algorithm 8 BUILDF_PH1($\mathcal{B}, \mathcal{S}$)—Build $\mathrm{F} \subseteq \mathrm{F}^{ph1}_{\mathcal{B}}$

Input: A f.c.b. $\mathcal{B}$ such that $dim(\mathcal{IR}^{ph1}_{\mathcal{B}}) = k+1$ and a set $\mathcal{S} \subseteq \mathcal{B}$.
Output: The set F.

1: Let $\mathrm{F} = \emptyset$.
2: **for** $i \in \left(\mathcal{S} \setminus \left(\mathrm{Z}^{ph1}_{\mathcal{B}} \cup \mathrm{E}^{ph1}_{\mathcal{B}} \cup \mathrm{F}\right)\right)$ **do** solve $NLP^{ph1}_{F}(\mathcal{B}, i)$ to find an optimal solution $(\lambda^*, \rho^*, \phi^*, \upsilon^*)$.
3: **if** $\lambda^* > 0$ **then** add $\left(i \cup (\mathrm{H}^{i}_{\mathcal{B}})^{ph1}\right)$ to F.
4: Return F.

Algorithm 9 GETADJACENTREGIONSACROSS_PH1($\mathcal{B}$,i,$\mathscr{B}$)—Determine all f.c.b.'s with associated invariancy regions that are adjacent to $\mathcal{IR}_{\mathcal{B}}^{ph1}$ across $(\hbar_{\mathcal{B}}^{i})^{ph1}$

Input: A f.c.b. $\mathcal{B}$, an index $i \in \mathrm{F}_{\mathcal{B}}$ and the set $\mathscr{B}$ of previously processed f.c.b.'s.
Output: The set $\mathcal{S}'$ of previously undiscovered f.c.b.'s with associated invariancy regions that are adjacent to $\mathcal{IR}_{\mathcal{B}}^{ph1}$ across $(\hbar_{\mathcal{B}}^{i})^{ph1}$, the set $\mathcal{S}''$ of previously discovered f.c.b.'s with associated invariancy regions that are potentially adjacent to $\mathcal{IR}_{\mathcal{B}}^{ph1}$ across $(\hbar_{\mathcal{B}}^{i})^{ph1}$, and an updated version of $\mathscr{B}$.

Set $\mathcal{S}' = \emptyset$, $\mathcal{S}'' = \emptyset$, and $\mathcal{B}' = \big(\mathcal{B} \setminus \{i\}\big) \cup \{\bar{\imath}\}$. Select an arbitrary $\phi' \in \Phi$.
if $\mathcal{B}' \notin \mathscr{B}$ **then**
  if $(T_{\mathcal{B}}^{ph1}(\phi))_{i,\bar{\imath}} \not\equiv 0$ **then** set $\mathscr{B} = \mathscr{B} \cup \{\mathcal{B}'\}$ and let d = GETINVARIANCYREGIONDIMENSION_PH1($\mathcal{B}'$).
    if $d = k + 1$ **then** set $\mathcal{S}' = \{\mathcal{B}'\}$.
    else set $(\mathcal{S}', \mathscr{B})$ = FINDADJACENTK_PH1($\mathcal{B}', \bar{\imath}, \mathcal{B}, i, \mathscr{B}$).
  else
    for $j \in \mathcal{B} \setminus \{i\}$ **do**
      Set $\mathcal{B}' = \big(\mathcal{B} \setminus \{i, j\}\big) \cup \{\bar{\imath}, \bar{\jmath}\}$
      if $\mathcal{B}' \notin \mathscr{B}$ and $(T_{\mathcal{B}}^{ph1}(\phi'))_{j,\bar{\imath}}$ either has degree greater of at least 1 or is a strictly positive constant **then** solve $NLP_{A}^{ph1}(\mathcal{B}, i, j)$ to obtain an optimal solution $(\lambda^*, \phi^*, \upsilon^*)$.
        if $\lambda^* > 0$ **then** set $\mathscr{B} = \mathscr{B} \cup \{\mathcal{B}'\}$ and let d = GETINVARIANCYREGIONDIMENSION_PH1($\mathcal{B}'$).
          if $d = k + 1$ **then** set $\mathcal{S}' = \mathcal{S}' \cup \{\mathcal{B}'\}$.
          else set $(\mathcal{S}, \mathscr{B})$ = FINDADJACENTK_PH1($\mathcal{B}', \bar{\jmath}, \mathcal{B}, i, \mathscr{B}$) and $\mathcal{S}' = \mathcal{S} \cup \mathcal{S}'$.
      else if $\mathcal{B}' \in \mathscr{B}$ **then** set $\mathcal{S}'' = \mathcal{S}'' \cup \mathcal{B}'$.
else set $\mathcal{S}'' = \mathcal{S}'' \cup \mathcal{B}'$.
Return $(\mathcal{S}', \mathcal{S}'', \mathscr{B})$.

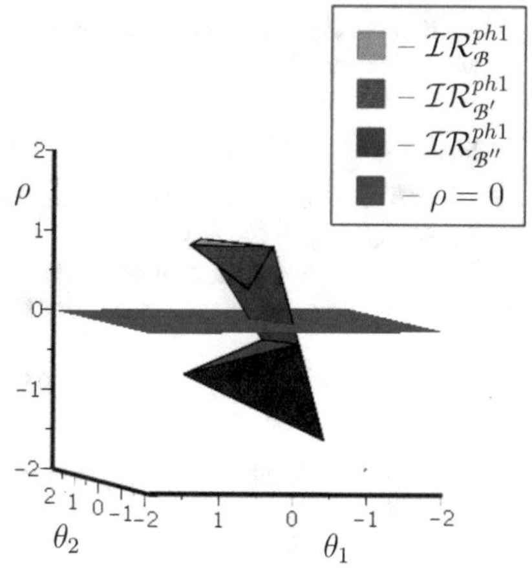

Fig. 5.1 Example of pivots yielding $(k+1)$-dimensional phase 1 regions on either side of $\rho = 0$, but without generating a full dimensional phase 2 region

that $\mathcal{IR}_{\mathcal{B}''}^{ph1}$ is $(k + 1)$-dimensional but is contained in the half-space $\rho \leq 0$ and $dim(\mathcal{IR}_{\mathcal{B}''}) < k$. See Figure 5.1 for a visual example of this situation with $k = 2$.

To ensure that this situation does not arise, when we encounter a phase 1 invariancy region $\mathcal{IR}_{\mathcal{B}}^{ph1}$ that is k-dimensional, we do something stronger than search for adjacent phase 1 invariancy regions. Instead, we search for phase 1 invariancy regions that are not only adjacent to $\mathcal{IR}_{\mathcal{B}}^{ph1}$ but are also adjacent to a full dimensional region that we know $\mathcal{IR}_{\mathcal{B}}^{ph1}$ to be adjacent to. Note that this full dimensional region is guaranteed to be available since phase 1 always begins with $\mathcal{B}^*$, which has a full dimensional region (see Proposition 5.1). For this reason, we use the third and fourth inputs to FINDADJACENTK_PH1 to keep track of (i) the f.c.b., say $\mathcal{B}'$, whose associated invariancy region is full dimensional and adjacent to $\mathcal{IR}_{\mathcal{B}}^{ph1}$ and (ii) an index $i \in \mathcal{B}'$ for which we know $\mathcal{IR}_{\mathcal{B}}^{ph1}$ is adjacent to $\mathcal{IR}_{\mathcal{B}'}^{ph1}$ along $(\hbar_{\mathcal{B}'}^{i})^{ph1}$. The routine FINDADJACENTK_PH1 is described in Algorithm 10.

Algorithm 10 FINDADJACENTK_PH1($\mathcal{B}_1$,ℓ_1,$\mathcal{B}_2$,ℓ_2,$\mathscr{B}$)—-Determine all previously undiscovered phase 1 invariancy regions that are adjacent to a given k-dimensional phase 1 invariancy region and a previously discovered $(k+1)$-dimensional phase 1 invariancy region

---

Input: A f.c.b. $\mathcal{B}_1$ such that $dim(\mathcal{IR}_{\mathcal{B}_1}^{ph1}) = k$, an index $\ell_1 \in \mathcal{B}_1$ for which $dim\left(\mathcal{IR}_{\mathcal{B}_1}^{ph1} \cap (\hbar_{\mathcal{B}_1}^{\ell_1})^{ph1}\right) = k$, a f.c.b. $\mathcal{B}_2$ such that $dim(\mathcal{IR}_{\mathcal{B}_1}^{ph1}) = k+1$ and $\mathcal{IR}_{\mathcal{B}_1}^{ph1}$ is adjacent to $\mathcal{IR}_{\mathcal{B}_2}^{ph1}$, an index $\ell_2 \in \mathcal{B}_2$ for which $\mathcal{IR}_{\mathcal{B}_1}^{ph1}$ is adjacent to $\mathcal{IR}_{\mathcal{B}_2}^{ph1}$ along $(\hbar_{\mathcal{B}_2}^{\ell_2})^{ph1}$, and the set $\mathscr{B}$ of previously discovered bases.

Output: The set $\mathcal{S}'$ of f.c.b.'s with associated invariancy regions that are adjacent to $\mathcal{IR}_{\mathcal{B}_1}$ and $\mathcal{IR}_{\mathcal{B}_2}$ and an updated version of $\mathscr{B}$.

Set $\mathcal{S}' = \emptyset$ and $(\mathrm{D}_{\mathcal{B}_1}^{\ell_1})^{ph1} = \emptyset$. Select an arbitrary $\phi' \in \Phi$.
for $j \in \mathcal{B}_1 \setminus \{\ell_1\}$ **do** compute $GCD(\mathcal{B}_1, \ell_1, j)$.
    if $GCD(\mathcal{B}_1, \ell_1, j)$ is a nonconstant polynomial **then** solve $NLP_G^{ph1}(\mathcal{B}_1, \ell_1, j)$ to obtain an optimal solution $(\lambda^*, \phi^*, \upsilon^*)$.
        if $\lambda^* = 0$ **then** add j to $\mathrm{D}_{\mathcal{B}_1}^{\ell_1}$.
for $i \in (\mathrm{D}_{\mathcal{B}_1}^{\ell_1})^{ph1}$ **do** set $\mathcal{B}' = (\mathcal{B}_1 \setminus \{i\}) \cup \{\bar{i}\}$.
    if $\mathcal{B}' \notin \mathscr{B}$ and $(T_{\mathcal{B}_1}^{ph1}(\phi))_{i,\bar{i}} \not\equiv 0$ **then** set $\mathscr{B} = \mathscr{B} \cup \{\mathcal{B}'\}$ and let $d =$ GETINVARIANCYREGIONDIMENSION_PH1($\mathcal{B}'$).
        if $d = k+1$ **then** set $\mathcal{S}' = \mathcal{S}' \cup \{\mathcal{B}'\}$.
        else set $(\mathcal{S}'', \mathscr{B}) =$ FINDADJACENTK_PH1($\mathcal{B}', \bar{i}, \mathcal{B}_2, \ell_2, \mathscr{B}$) and $\mathcal{S}' = \mathcal{S}' \cup \mathcal{S}''$.
    else
        for $j \in \mathcal{B}_1 \setminus \{i\}$ **do** set $\mathcal{B}'' = (\mathcal{B}_1 \setminus \{i, j\}) \cup \{\bar{i}, \bar{j}\}$.
            if $\mathcal{B}'' \notin \mathscr{B}$ and $(T_{\mathcal{B}_1}^{ph1}(\phi'))_{j,\bar{i}}$ either has degree greater than 1 or is a strictly positive constant **then** compute $T_{\mathcal{B}''}^{ph1}(\phi)$ and solve $NLP_{A2}^{ph1}(\mathcal{B}_1, i, j, \mathcal{B}_2, \ell_2)$ to obtain an optimal solution $(\lambda^*, \phi^*, \upsilon^*)$.
                if $\lambda^* > 0$ **then** set $\mathscr{B} = \mathscr{B} \cup \{\mathcal{B}''\}$ and let $d =$ GETINVARIANCYREGIONDIMENSION($\mathcal{B}''$).
                    if $d = k+1$ **then** set $\mathcal{S}' = \mathcal{S}' \cup \{\mathcal{B}''\}$.
                    else set $(\mathcal{S}'', \mathscr{B}) =$ FINDADJACENTK_PH1($\mathcal{B}'', \bar{j}, \mathcal{B}_2, \ell_2, \mathscr{B}$) and $\mathcal{S}' = \mathcal{S}' \cup \mathcal{S}''$.
Return $\mathcal{S}'$ and $\mathscr{B}$.

Recognize from Algorithm 10 that the routine FINDADJACENTK_PH1 is quite similar to its phase 2 counterpart, FINDADJACENTKMINUS1 (Algorithm 6). The only difference between these two routines comes from the fact that in order to ensure that each f.c.b. discovered in Algorithm 10 has an invariancy region that is adjacent to both $\mathcal{IR}_{\mathcal{B}_1}$ and $\mathcal{IR}_{\mathcal{B}_2}$, the structure of NLP^{ph1}_{A2} must be modified from the structure of NLP_{A2}. Evidence of this modification can be seen on line 11 of Algorithm 10 because NLP^{ph1}_{A2} takes five inputs, whereas NLP_{A2} only takes three. The specifics of NLP^{ph1}_{A2} are presented in the following proposition, which serves as the phase 1 counterpart to Proposition 4.13.

Proposition 5.4 *Let distinct f.c.b.'s $\mathcal{B}_1$, $\mathcal{B}_2$, and $\mathcal{B}'$ be given for which $dim(\mathcal{IR}^{ph1}_{\mathcal{B}_1}) = k$, $dim(\mathcal{IR}^{ph1}_{\mathcal{B}_2}) = k+1$, $\mathcal{IR}^{ph1}_{\mathcal{B}_1}$ is adjacent to $\mathcal{IR}^{ph1}_{\mathcal{B}_2}$, and $|\mathcal{B} \cap \mathcal{B}'| \geq h-2$. Furthermore, let $\ell_1 \in \mathcal{B}_1$ be an index for which $dim\left(\mathcal{IR}^{ph1}_{\mathcal{B}_1} \cap (\mathcal{h}^{\ell_1}_{\mathcal{B}_1})^{ph1}\right) = k$, and let $\ell_2 \in \mathcal{B}_2$ be the index for which we know $\mathcal{IR}^{ph1}_{\mathcal{B}_1}$ is adjacent to $\mathcal{IR}^{ph1}_{\mathcal{B}_2}$ along $(\mathcal{h}^{\ell_2}_{\mathcal{B}_2})^{ph1}$. Then $\mathcal{IR}^{ph1}_{\mathcal{B}'}$ is adjacent to $\mathcal{IR}^{ph1}_{\mathcal{B}_2}$ along $(\mathcal{h}^{\ell_1}_{\mathcal{B}_1})^{ph1}$ if and only if there exists $i \in (D^{\ell_1}_{\mathcal{B}_1})^{ph1}$ such that one of the following conditions holds:*

1. *$\mathcal{B}' = (\mathcal{B}_1 \setminus \{i\}) \cup \{\bar{\imath}\}$ and $(T^{ph1}_{\mathcal{B}_1}(\phi))_{i,\bar{\imath}} \not\equiv 0$.*
2. *There exists an index $j \in \mathcal{B}_1 \setminus \{i\}$ such that $\mathcal{B}' = (\mathcal{B}_1 \setminus \{i, j\}) \cup \{\bar{\imath}, \bar{\jmath}\}$, $(T^{ph1}_{\mathcal{B}_1}(\phi))_{i,\bar{\imath}} \equiv 0$, and the following NLP has a strictly positive optimal value:*

$$
\begin{aligned}
&NLP^{ph1}_{A2}(\mathcal{B}_1, i, j, \mathcal{B}_2, \ell_2) := \\
&\max_{\lambda,\phi,\upsilon} \quad \lambda \\
&\text{s.t.} \quad g_{\mathcal{B}_2}\left(Adj(G(\phi)_{\cdot\mathcal{B}_2})\right)_{j\cdot} G(\phi)_{\cdot\bar{\ell_2}} \geq \lambda \\
&\quad g_{\mathcal{B}_2}\left(Adj(G(\phi)_{\cdot\mathcal{B}_2})\right)_{\xi\cdot}(q(\phi,\upsilon)+r\rho) \geq \lambda \quad \forall \xi \in \left(\mathcal{B}_2 \setminus \left(Z^{ph1}_{\mathcal{B}_2} \cup (H^{\ell_2}_{\mathcal{B}_2})^{ph1} \cup \{\ell_2\}\right)\right) \\
&\quad \left(Adj(G(\phi)_{\cdot\mathcal{B}_2})\right)_{\ell_2\cdot}(q(\phi,\upsilon)+r\rho) = 0 \\
&\quad g_{\mathcal{B}'}\left(Adj(G(\phi)_{\cdot\mathcal{B}'})\right)_{\xi\cdot}(q(\phi,\upsilon)+r\rho) \geq \lambda \quad \forall \xi \in \left(\mathcal{B}' \setminus \left(Z^{ph1}_{\mathcal{B}'} \cup (H^{\bar{\jmath}}_{\mathcal{B}'})^{ph1} \cup \{\bar{\jmath}\}\right)\right) \\
&\quad \phi \in \Phi, \upsilon \in \Upsilon
\end{aligned}
\tag{5.6}
$$

Proof It is clear from analogous arguments to those given in Chapter 4, namely Proposition 4.13, together with Observations 4.6, (4.20), (4.21), and (4.22), that $dim\left(\mathcal{IR}^{ph1}_{\mathcal{B}_1} \cap (\mathcal{h}^{\ell_1}_{\mathcal{B}_1})^{ph1} \cap (\mathcal{h}^{i}_{\mathcal{B}_1})^{ph1}\right) = k$ if and only if $i \in (D^{\ell}_{\mathcal{B}})^{ph1}$. Furthermore, it is also clear that if $\mathcal{B}_1$ was obtained from $\mathcal{B}_2$ by a single diagonal pivot as prescribed in condition (1) or a single exchange pivot as prescribed in condition (2), then $dim\left(\mathcal{IR}^{ph1}_{\mathcal{B}_1} \cap (\mathcal{h}^{\ell_1}_{\mathcal{B}_1})^{ph1} \cap (\mathcal{h}^{\ell_2}_{\mathcal{B}_2})^{ph1}\right) = k$. We also have $dim\left(\mathcal{IR}^{ph1}_{\mathcal{B}_1} \cap (\mathcal{h}^{\ell_1}_{\mathcal{B}_1})^{ph1} \cap (\mathcal{h}^{i}_{\mathcal{B}_1})^{ph1} \cap (\mathcal{h}^{\ell_2}_{\mathcal{B}_2})^{ph1}\right) = k$. Now suppose that $\mathcal{B}_1$ was obtained from $\mathcal{B}_2$ by a sequence of pivots to intermediate bases, with each pivot satisfying either condition (1) or condition (2). In the case of pivots satisfying condition (1), the adjacency of each newly discovered region to $\mathcal{IR}_{\mathcal{B}_2}$ along $(\hbar_{\mathcal{B}_1}^{\ell_1})^{ph1}$ is clear due to the uniqueness of shared k-dimensional boundaries for regions obtained from diagonal pivots. In the case of pivots satisfying condition (2), recognize from (5.6) that the second set of constraints of NLP_{A2}^{ph1} ensures that at any solution for which $\lambda > 0$, all defining constraints of $\mathcal{IR}_{\mathcal{B}_2}^{ph1}$ are satisfied strictly except for those implied by $(\hbar_{\mathcal{B}_1}^{\ell_1})^{ph1}$. The second constraint ensures that solutions lie on $(\hbar_{\mathcal{B}_1}^{\ell_1})^{ph1}$. Finally, the third set of constraints ensures that at any solution for which $\lambda > 0$, all defining constraints of the newly discovered invariancy region are satisfied strictly except for those implied by $(\hbar_{\mathcal{B}_1}^{\ell_1})^{ph1}$. Recognize that this can only hold if $dim\left(\mathcal{IR}_{\mathcal{B}_1}^{ph1} \cap (\hbar_{\mathcal{B}_1}^{\ell_1})^{ph1} \cap (\hbar_{\mathcal{B}_1}^{i})^{ph1} \cap (\hbar_{\mathcal{B}_2}^{\ell_2})^{ph1}\right) = k$. Hence, we conclude that $\mathcal{IR}_{\mathcal{B}_2}^{ph1}$ and $\mathcal{IR}_{\mathcal{B}_1}^{ph1}$ are adjacent along $(\hbar_{\mathcal{B}_1}^{\ell_1})^{ph1}$ regardless of the number of pivots taken to obtain $\mathcal{B}_1$ from $\mathcal{B}_2$, so long as each pivot satisfies either condition (1) or condition (2). The remainder of the proof is analogous to those of Propositions 4.7 and 4.13. □

We now use the following proposition to argue the correctness of Algorithm 7.

Proposition 5.5 *Given an instance of mpLCP as described in* (1.1), *if* $M(\phi)$ *is sufficient for all* ϕ *in* Φ *and there exists a f.c.b.* $\mathcal{B}$ *such that* $dim(\mathcal{IR}_{\mathcal{B}}) = k$, *Algorithm 7 will return a f.c.b.* $\mathcal{B}'$ *such that* $dim(\mathcal{IR}_{\mathcal{B}'}) = k$.

Proof Recognize that Algorithm 7 only ceases if (i) the Return command is reached on line 5 or (ii) the STOP command on either line 14 or line 15 is reached. Clearly if the Return command on line 5 is reached, the claim of the proposition holds since a full dimensional phase 2 invariancy region is returned. On the other hand, if the STOP command on line 14 is reached, the claim of the proposition is also satisfied since the correctness of this stopping criterion is proved in Proposition 5.3. Now, note that the STOP command on line 15 is only reached if there are no more bases in $\mathcal{S}$ to explore. The theory we have developed in this chapter ensures that, given a f.c.b. $\mathcal{B}$ discovered during phase 1 for which $dim(\mathcal{IR}_{\mathcal{B}}) \neq k$, the only indices in $\mathcal{B}$ for which we do not consider performing pivots are precisely those that have no possibility of yielding a new basis $\mathcal{B}'$ such that $dim(\mathcal{IR}_{\mathcal{B}'}) = k$. Hence, the result of the proposition holds in this case as well. □

We now return to Example 2.1 and briefly discuss how the theory presented in this chapter can be used to obtain an initial basis. We omit the consideration of phase 1 for Example 2.2 because only one iteration is necessary. It is easy to verify that the optimal value of $NLP_S(\mathcal{B}^{*\,2.2})$ (5.4) is nonpositive and the optimal value of $NLP_D(\mathcal{B}^{*\,2.2})$ (4.19) is strictly positive, and so $\mathcal{B}^{*\,2.2}$ serves as the initial basis for Example 2.2. Now, from Table A.1 in Appendix A, we construct $NLP_S(\mathcal{B}^{*\,2.1})$:

$$
\begin{aligned}
\min_{\phi,\rho} \quad & \rho \\
\text{s.t.} \quad & 3 \geq 0 \\
& -2 - \phi_1 + 3\rho \geq 0 \\
& 0 \geq 0 \\
& 0 \geq 0 \\
& 0 \geq 0 \\
& \phi_1 + \phi_2 \leq 1 \\
& \phi_1, \phi_2 \geq 0
\end{aligned}
$$

An approximate optimal solution is $(\phi_1^*, \phi_2^*, \rho^*) = (0, 0, 0.6667)$. This shows that $\mathcal{IR}^{ph1}_{\mathcal{B}^{*2.1}}$ does not intersect the hyperplane $\rho = 0$, and thus $\mathcal{B}^{*2.1}$ is infeasible for mpLCP. From (5.5) we find that $EQ_{\mathcal{B}^{*2.1}} = \{w_2, w_3, w_4, w_5\}$. However, recognize that $\mathrm{Z}^{ph1}_{\mathcal{B}^{*2.1}} = \{w_3, w_4, w_5\}$. Hence, w_2 is the only variable in $\mathcal{IR}^{ph1}_{\mathcal{B}^{*2.1}}$ that is considered during the call to BUILDF_PH1. The optimal value of $NLP_F^{ph1}(\mathcal{B}^{*2.1})$ (see (4.7) for NLP_F) is unbounded, which shows that $(\hbar^{w_2}_{\mathcal{B}^{*2.1}})^{ph1}$ forms a k-dimensional boundary of $\mathcal{IR}^{ph1}_{\mathcal{B}^{*2.1}}$. The next step is then to call GETADJACENTREGIONSACROSS_PH1($\mathcal{B}^{*2.1}, w_2, \mathscr{B} = \{\mathcal{B}^{*2.1}\}$). It can be observed from $T^{ph1}_{\mathcal{B}^{*2.1}}(\phi)$, found in Table A.1 in Appendix A, that although a diagonal pivot from w_2 is not possible, exchange pivots involving w_2 along with w_3, w_4, or w_5 are possible. We consider these one at a time, beginning with w_3. An approximate optimal solution of $NLP_A^{ph1}(\mathcal{B}^{*2.1}, w_2, w_3)$, given in the order $(\lambda, \rho, \phi_1, \phi_2)$, is $(2, 0.7071, 0.1212, 0)$. This shows that basis $\mathcal{B}_i^{2.1} = \{w_1, z_2, z_3, w_4, w_5\}$ yields a phase 1 invariancy region adjacent to $\mathcal{IR}^{ph1}_{\mathcal{B}^{*2.1}}$ across $(\hbar^{w_2}_{\mathcal{B}^{*2.1}})^{ph1}$ (see Table A.2 for $T^{ph1}_{\mathcal{B}_i^{2.1}}(\phi, \rho)$). We now proceed to call GETINVARIANCYREGIONDIMENSION_PH1($\mathcal{B}_i^{2.1}$). An approximate optimal solution of $NLP_D^{ph1}(\mathcal{B}_i^{2.1})$ (see (4.19) for NLP_D) is $(0, 0, 0, 0.6667)$, which shows that $\mathcal{IR}^{ph1}_{\mathcal{B}_i^{2.1}}$ is not full dimensional. Thus, we run FINDADJACENTK_PH1($\mathcal{B}_i^{2.1}, z_3, \mathcal{B}^{*2.1}, w_2, \mathscr{B} = \{\mathcal{B}^{*2.1}, \mathcal{B}_i^{2.1}\}$). In doing so, we find $GCD^{ph1}(\mathcal{B}_i^{2.1}, z_3, w_1) = \frac{1}{2}$, $GCD^{ph1}(\mathcal{B}_i^{2.1}, z_3, z_2) = \frac{1}{4}\phi_1 - \frac{3}{4}\rho + \frac{1}{2}$, and $GCD^{ph1}(\mathcal{B}_i^{2.1}, z_3, w_4) = GCD^{ph1}(\mathcal{B}_i^{2.1}, z_3, w_5) = \frac{1}{2}\phi_1 - \frac{3}{2}\rho + 1$. We also find that the optimal values of $NLP_G^{ph1}(\mathcal{B}_i^{2.1}, z_3, z_2)$, $NLP_G^{ph1}(\mathcal{B}_i^{2.1}, z_3, w_4)$, and $NLP_G^{ph1}(\mathcal{B}_i^{2.1}, z_3, w_5)$ (see (4.23) for NLP_G) are all zero, so we have $(\mathrm{D}^{z_3}_{\mathcal{B}_i^{2.1}})^{ph1} = \{z_2, w_4, w_5\}$. We observe from Table A.2 in Appendix A that for each variable in $(\mathrm{D}^{z_3}_{\mathcal{B}_i^{2.1}})^{ph1}$, a diagonal pivot is possible. Hence, we first perform the diagonal pivot on z_2, which yields $\mathcal{B}_{ii}^{2.1} = \{w_1, w_2, z_3, w_4, w_5\}$ (see Table A.3 for $T^{ph1}_{\mathcal{B}_{ii}^{2.1}}(\phi, \rho)$). Calling GETINVARIANCYREGIONDIMENSION_PH1($\mathcal{B}_{ii}^{2.1}$) shows that $\mathcal{IR}^{ph1}_{\mathcal{B}_{ii}^{2.1}}$ is $(k+1)$-dimensional, so we add $\mathcal{B}_{ii}^{2.1}$ to $\mathcal{S}'$ and proceed to consideration of w_4. The diagonal pivot on w_4 yields $\mathcal{B}_{iii}^{2.1} = \{w_1, z_2, z_3, z_4, w_5\}$, which has a k-dimensional phase 1 invariancy region (see Table A.4 for $T^{ph1}_{\mathcal{B}_{iii}^{2.1}}(\phi, \rho)$). Thus, we must call FINDADJACENTK_PH1($\mathcal{B}_{iii}^{2.1}, z_4, \mathcal{B}^{*\,2.1}, w_2, \mathscr{B} = \{\mathcal{B}^{*\,2.1}, \mathcal{B}_i^{2.1}, \mathcal{B}_{ii}^{2.1}, \mathcal{B}_{iii}^{2.1}\}$). We do not include all of the details of this call, as the steps are analogous to those we are now considering, but we do point out that the sets returned from this call are $\mathcal{S}'' = \{\mathcal{B}_v^{2.1} = \{w_1, z_2, w_3, z_4, z_5\}, \mathcal{B}_{vi}^{2.1} = \{w_1, z_2, z_3, z_4, z_5\}\}$, and $\mathscr{B} = \{\mathcal{B}^{*\,2.1}, \mathcal{B}_i^{2.1}, \mathcal{B}_{ii}^{2.1}, \mathcal{B}_{iii}^{2.1}, \mathcal{B}_{iv}^{2.1} = \{w_1, z_2, w_3, z_4, w_5\}, \mathcal{B}_v^{2.1}, \mathcal{B}_{vi}^{2.1}\}$ ($T^{ph1}_{\mathcal{B}_{iv}^{2.1}}(\phi, \rho)$, $T^{ph1}_{\mathcal{B}_v^{2.1}}(\phi, \rho)$, and $T^{ph1}_{\mathcal{B}_{vi}^{2.1}}(\phi, \rho)$ can be found in Tables A.5, A.6, and A.7, respectively). We also note that the call to FINDADJACENTK_PH1($\mathcal{B}_{iii}^{2.1}, z_4, \mathcal{B}^{*\,2.1}, w_2, \mathscr{B} = \{\mathcal{B}^{*\,2.1}, \mathcal{B}_i^{2.1}, \mathcal{B}_{ii}^{2.1}, \mathcal{B}_{iii}^{2.1}\}$) requires one additional level of recursion because the discovered basis $\mathcal{B}_{iv}^{2.1}$ does not have a full dimensional phase 1 invariancy region. We now set $\mathcal{S}' = \mathcal{S}' \cup \mathcal{S}'' = \{\mathcal{B}_{ii}^{2.1}, \mathcal{B}_v^{2.1}, \mathcal{B}_{vi}^{2.1}\}$ and return to our consideration of $\mathcal{B}_i^{2.1}$ and the diagonal pivot on w_5. Performing this pivot yields $\mathcal{B}_{vii}^{2.1} = \{w_1, z_2, z_3, w_4, z_5\}$, which does not have a full dimensional phase 1 invariancy region (see Table A.8 for $T^{ph1}_{\mathcal{B}_{vii}^{2.1}}(\phi, \rho)$). As a result, we must call FINDADJACENTK_PH1($\mathcal{B}_{vii}^{2.1}, z_5, \mathcal{B}^{*\,2.1}, w_2, \mathscr{B} = \{\mathcal{B}^{*\,2.1}, \mathcal{B}_i^{2.1}, \ldots, \mathcal{B}_{vii}^{2.1}\}$). Again, we omit the details of this call, but note that the returned sets are $\mathcal{S}'' = \emptyset$ and $\mathscr{B} = \{\mathcal{B}^{*\,2.1}, \mathcal{B}_i^{2.1}, \ldots, \mathcal{B}_{viii}^{2.1} = \{w_1, z_2, w_3, w_4, z_5\}\}$ (see Table A.9 for $T^{ph1}_{\mathcal{B}_{viii}^{2.1}}(\phi, \rho)$). Having now completed the call to FINDADJACENTK_PH1($\mathcal{B}_i^{2.1}, z_3, \mathcal{B}^{*\,2.1}, w_2, \mathscr{B}$), we return the sets $\mathcal{S}'$ and $\mathscr{B}$, thus arriving back in our original consideration of the call to GETADJACENTREGIONSACROSS_PH1($\mathcal{B}^{*\,2.1}, w_2, \mathscr{B}$). Here, the final step is to reconsider the exchange pivots from $\mathcal{B}^{*\,2.1}$ involving w_2 and w_4, and w_2 and w_5. However, observe that the former pivot results in the rediscovery of $\mathcal{B}_{iv}^{2.1}$ and that the latter results in the rediscovery of $\mathcal{B}_{viii}^{2.1}$. Hence, we now exit GETADJACENTREGIONSACROSS_PH1 by returning the sets $\mathcal{S}' = \{\mathcal{B}_{ii}^{2.1}, \mathcal{B}_v^{2.1}, \mathcal{B}_{vi}^{2.1}\}$ and $\mathscr{B} = \{\mathcal{B}^{*\,2.1}, \mathcal{B}_i^{2.1}, \ldots, \mathcal{B}_{viii}^{2.1}\}$ and arrive back in the FINDINITIALBASIS routine. We cease our consideration of Example 2.1 here because the next basis considered in FINDINITIALBASIS is $\mathcal{B}_{vi}^{2.1}$, which not only has a full dimensional phase 1 invariancy region but also has a full dimensional phase 2 invariancy region. Thus, $\mathcal{B}_{vi}^{2.1}$ serves as the initial f.c.b. $\mathcal{B}_0$ with which we begin phase 2.

A careful analysis of the work performed above reveals an interesting phenomenon, namely, $\mathcal{IR}^{ph1}_{\mathcal{B}_{ii}^{2.1}} = \mathcal{IR}^{ph1}_{\mathcal{B}^{*\,2.1}}$. This shows that invariancy regions, both in phase 1 and in phase 2, can overlap. Moreover, this implies that not every instance of mpLCP has a unique partition of Θ. We provide a discussion on this phenomenon and its implications in Section 6.2.

Chapter 6
Further Considerations

The purpose of this chapter is to consider difficulties that result from the structure of certain instances of mpLCP and specify the methods we employ in order to properly manage these difficulties. In Section 6.1 we offer examples of instances of mpLCP that motivate Assumptions 1.1 and 1.2. Specifically, we address potential consequences of *not* satisfying these assumptions. In Section 6.2 we consider the uniqueness of partitions of Θ. In particular, we discuss conditions under which two invariancy regions will overlap, i.e., have a k-dimensional intersection.

6.1 On the Importance of Assumptions 1.1 and 1.2

We now introduce the reader to two additional examples that highlight the importance of Assumptions 1.1 and 1.2 (and, by extension, Assumption 5.1). The first example is constructed so that Assumption 1.1 is satisfied and we discuss its solution over two variants of Θ, one in which Assumption 1.2 is satisfied, and one in which it is not. The second example is constructed so that Assumption 1.1 is not satisfied and demonstrate how the methodology presented in this work breaks down in this case. We now present the first of these examples.

Example 6.1

$$\begin{aligned} w - \begin{bmatrix} 0 & \theta_1 - 1 \\ -\theta_1 + 1 & 0 \end{bmatrix} z = \begin{bmatrix} \theta_1 \\ \theta_2 \end{bmatrix} \\ w^\top z = 0 \\ w, z \geq 0 \end{aligned} \tag{6.1}$$

N. Adelgren, *Advancing Parametric Optimization*, SpringerBriefs in Optimization,
https://doi.org/10.1007/978-3-030-61821-6_6

Assume for now that $\Theta = [-2, 2] \times [-2, 2]$ and notice that $M(\theta) = \begin{bmatrix} 0 & \theta_1 - 1 \\ -\theta_1 + 1 & 0 \end{bmatrix}$ is sufficient for all $\theta \in \Theta$ by Lemma 1.1. It can be easily verified that the initial basis $\mathcal{B}_0 = \{w_1, w_2\}$ has full dimensional invariancy region $\mathcal{IR}_{\mathcal{B}_0} = \{\theta \in \Theta : \theta_1 \geq 0, \theta_2 \geq 0\}$. This region can be seen in red in Figure 6.1a. It is also easy to verify that a single exchange pivot yields the only other f.c.b. for this problem, $\mathcal{B}_1 = \{z_1, z_2\}$. We note the associated invariancy region is $\mathcal{IR}_{\mathcal{B}_1} = \{\theta \in \Theta : -\theta_1(\theta_1 - 1) \geq 0, \theta_2(\theta_1 - 1) \geq 0, \theta_1 \neq 1\}$ and is depicted in yellow in Figure 6.1a.

Suppose now that we replace Θ with $\Theta' = \{\theta \in \mathbb{R}^2 : \theta_1 \leq 2, \theta_2 \geq -2, -5\theta_1 + 3\theta_2 \leq -5\}$ as depicted in Figure 6.1b. As $\Theta' \subset \Theta$, we know $M(\theta)$ is still sufficient for all $\theta \in \Theta'$. From this scenario, we make two important observations.

Observation 6.1 *Even in the case in which $M(\theta)$ is sufficient for all $\theta \in \Theta$, the union of invariancy regions is not necessarily convex.*

Observation 6.2 *Even in the case in which $M(\theta)$ is sufficient for all $\theta \in \Theta$, the set $\hat{\Theta}$ (1.2) is not necessarily connected.*

We note that Observation 6.2 motivates our need for Assumption 1.2, as without it the methodology presented in this work cannot be guaranteed to discover a solution to mpLCP (1.1) for every feasible $\theta \in \Theta$. We note that one potential way to avoid this dilemma is to force mpLCP (1.1) to be constructed so that there exists $\theta' \in relint(\Theta)$ such that $q(\theta') = \mathbf{0}$. Recognize that if this is not the case for a given instance of mpLCP, it can be achieved by adding an additional parameter and/or enlarging Θ.

So now suppose there exists $\theta' \in relint(\Theta)$ such that $q(\theta') = \mathbf{0}$ and recall that every nonempty parametric complementary cone contains the origin. This has two important implications: (i) every complementary basis whose associated parametric complementary cone is nonempty is a f.c.b., (ii) and every invariancy region contains

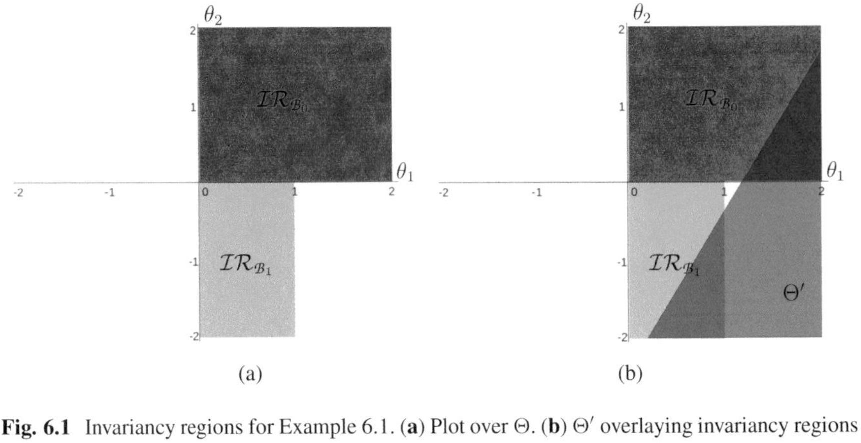

Fig. 6.1 Invariancy regions for Example 6.1. (**a**) Plot over Θ. (**b**) Θ' overlaying invariancy regions

θ'. Thus, whenever Assumption 1.1 is satisfied, the methodology presented herein is capable of discovering every full dimensional invariancy region. We point out, however, that for reasonably sized instances of mpLCP such a workaround may be very well be intractable as it might significantly increase the number of feasible bases considered and, as is well known, the number of possible feasible bases is exponential in the size of the instance.

Now consider another example that highlights the importance of Assumption 1.1.

Example 6.2

$$\begin{aligned} w - \begin{bmatrix} 0 & \theta_1 \\ \theta_1 & 0 \end{bmatrix} z &= \begin{bmatrix} \theta_1 \\ \theta_2 \end{bmatrix} \\ w^\top z &= 0 \\ w, z &\geq 0 \end{aligned} \tag{6.2}$$

As it is unimportant for this example, we simply assume $\Theta = \mathbb{R}^2$. Observe Lemma 1.1 and recognize that in this case $M(\theta)$ cannot be sufficient for any $\theta \in \Theta$. Thus, rather than observing invariancy regions, we consider complementary cones in order to discover the difficulties that arise for this example. Recognize that for this example $h = 2$ and we therefore have $2^h = 4$ complementary cones. One each for: (i) the two identity columns, (ii) the first identity column and the second column of $-M(\theta)$, (iii) the first column of $-M(\theta)$ and the second identity column, and (iv) the two columns of $-M(\theta)$. First consider the case in which $\theta_1 < 0$. Viewing $\mathbb{R}^2$ in the traditional Euclidean coordinate system, the respective complementary cones are given by (i) quadrant 1, (ii) the nonnegative portion of the x-axis, (iii) the nonnegative portion of the y-axis, and (iv) quadrant 1. Recognize that these cones are not disjoint in their relative interiors. In fact, cones (i) and (iv) are identical. This means that it is far more likely that for a given $\theta \in \Theta$, we will have $q(\theta)$ present in more than one complementary cone. While this issue did arise before, it only occurred when $q(\theta)$ was on the boundary of a complementary cone, not in its relative interior. Now consider the case in which $\theta_1 > 0$. The respective complementary cones are now given by (i) quadrant 1, (ii) the x-axis, (iii) the y-axis, and (iv) quadrant 3. In this case, we see that the complementary cones are disjoint in their relative interiors, but unfortunately their union is not convex. This issue is surely more problematic than having cones whose relative interiors are not disjoint since the nonconvexity of the union of the cones ensures that the methodology presented herein can no longer be guaranteed to determine a solution to mpLCP for each feasible $\theta \in \Theta$ because we can no longer exploit the adjacency of complementary cones.

Having now discussed the difficulties that arise when Assumptions 1.1 and 1.2 are not satisfied, we now shift our focus to an issue that may emerge even when these assumptions are met. Namely, we consider distinct invariancy regions having k–dimensional intersection.

6.2 On Obtaining Non-overlapping Invariancy Regions

It is important to recognize that a partition of Θ for mpLCP is not unique, in general. Hence, if one is not careful, when attempting to partition Θ it is possible to generate invariancy regions $\mathcal{IR}_{\mathcal{B}}$ and $\mathcal{IR}_{\mathcal{B}'}$, associated with distinct complementary bases $\mathcal{B}$ and $\mathcal{B}'$, such that $dim(\mathcal{IR}_{\mathcal{B}} \cap \mathcal{IR}_{\mathcal{B}'}) = k$. In fact, this very situation arose during our phase 1 examination of Example 2.1. In this section we pose two important research questions, which we present directly after the following two definitions.

Definition 6.1 Given distinct complementary bases $\mathcal{B}$ and $\mathcal{B}'$, we say that $\mathcal{IR}_{\mathcal{B}}$ and $\mathcal{IR}_{\mathcal{B}'}$ *overlap* if $dim(\mathcal{IR}_{\mathcal{B}} \cap \mathcal{IR}_{\mathcal{B}'}) = k$. Otherwise, they are *non-overlapping*.

Definition 6.2 Given a feasible instance of mpLCP, let $\mathcal{P}$ be the output of Algorithm 1. We say that $\mathcal{P}$ is a *true partition* of Θ if no two invariancy regions in $\mathcal{P}$ overlap. Otherwise, we say that $\mathcal{P}$ is a *quasi-partition*.

Now consider these two important research questions.

Question 6.1 *Given a feasible instance of mpLCP, what conditions are needed to ensure that a true partition of Θ exists?*

Question 6.2 *When a true partition of Θ exists, what precautions can be taken to ensure that one is obtained?*

The remainder of this section is devoted to the consideration of these questions. We now introduce propositions whose results will allow us to establish an answer to Question 6.1.

Lemma 6.1 *Given distinct complementary bases $\mathcal{B}$ and $\mathcal{B}'$, the invariancy regions $\mathcal{IR}_{\mathcal{B}}$ and $\mathcal{IR}_{\mathcal{B}'}$ overlap if and only if there exists $\Phi' \subseteq \Phi$ such that $dim(\Phi') = p$ and $dim\left(\bigcup_{\upsilon \in \Upsilon} q(\phi, \upsilon) \cap \mathcal{C}_{\mathcal{B}}(\phi) \cap \mathcal{C}_{\mathcal{B}'}(\phi)\right) = k - p$ for all $\phi \in \Phi'$.*

Proof Recognize that the set $\mathcal{IR}_{\mathcal{B}} \cap \mathcal{IR}_{\mathcal{B}'}$ satisfies Property 3.1. Thus, by observing (3.6), (3.10) and Proposition 3.1 we conclude that the result of this lemma follows directly from Proposition 3.9. □

Proposition 6.1 *Let distinct complementary bases $\mathcal{B}$ and $\mathcal{B}'$ be given such that the invariancy regions $\mathcal{IR}_{\mathcal{B}}$ and $\mathcal{IR}_{\mathcal{B}'}$ overlap. Then there exist index sets $J \subset \mathcal{B}$ and $J' \subset \mathcal{B}'$ such that the following hold:*

1. $k - p \leq |J| \leq k - p + 1$ *and* $k - p \leq |J'| \leq k - p + 1$.
2. $dim(cone(G(\phi)_{\cdot J}) \cap cone(G(\phi)_{\cdot J'})) \geq k - p$.
3. $aff\left(\bigcup_{\upsilon \in \Upsilon} q(\phi, \upsilon)\right) \subseteq span(G(\phi)_{\cdot J^*})$ *for all* $\phi \in \Phi$*, where* $aff\left(\bigcup_{\upsilon \in \Upsilon} q(\phi, \upsilon)\right)$ *represents the affine hull of* $\bigcup_{\upsilon \in \Upsilon} q(\phi, \upsilon)$ *and* $J^* = \begin{cases} J & \text{if } |J| \geq |J'| \\ J' & \text{otherwise} \end{cases}$.

Proof From Lemma 6.1 we have that there exists $\Phi' \subseteq \Phi$ such that $dim(\Phi') = p$ and $dim\left(\bigcup_{\upsilon \in \Upsilon} q(\phi, \upsilon) \cap \mathcal{C}_{\mathcal{B}}(\phi) \cap \mathcal{C}_{\mathcal{B}'}(\phi)\right) = k - p$ for all $\phi \in \Phi'$. This shows that for each $\phi \in \Phi'$ the intersections $\bigcup_{\upsilon \in \Upsilon} q(\phi, \upsilon) \cap \mathcal{C}_{\mathcal{B}}(\phi)$ and $\bigcup_{\upsilon \in \Upsilon} q(\phi, \upsilon) \cap \mathcal{C}_{\mathcal{B}'}(\phi)$ must be contained within the boundaries of the respective parametric complementary cones. Furthermore, since for all $\phi \in \Phi$ we have $dim\left(\bigcup_{\upsilon \in \Upsilon} q(\phi, \upsilon)\right) = k - p$, for each $\phi \in \Phi$ and each $\hat{\mathcal{B}} \in \{\mathcal{B}, \mathcal{B}'\}$ the intersection $\bigcup_{\upsilon \in \Upsilon} q(\phi, \upsilon) \cap \mathcal{C}_{\hat{\mathcal{B}}}(\phi)$ must occur either within a $(k-p)$-dimensional facet of $\mathcal{C}_{\hat{\mathcal{B}}}(\phi)$, or within the convex hull of two $(k-p)$-dimensional facets (and therefore within a $(k-p+1)$-dimensional facet). For each $\phi \in \Phi'$, let $F_{\mathcal{B}}(\phi)$ and $F_{\mathcal{B}'}(\phi)$ respectively denote the facets of minimal dimension that contain $\bigcup_{\upsilon \in \Upsilon} q(\phi, \upsilon) \cap \mathcal{C}_{\mathcal{B}}(\phi)$ and $\bigcup_{\upsilon \in \Upsilon} q(\phi, \upsilon) \cap \mathcal{C}_{\mathcal{B}'}(\phi)$. Recall that for any basis $\mathcal{B}^*$ and any $\phi^* \in \Phi$, every ℓ-dimensional facet of $\mathcal{C}_{\mathcal{B}^*}(\phi^*)$ is given by $cone(G(\phi^*)_{\bullet I})$ for some $I \subseteq \mathcal{B}^*$ with $|I| = \ell$. Hence, we have that there exist $J \subset \mathcal{B}$ and $J' \subset \mathcal{B}'$ such that $cone(G(\phi)_{\bullet J}) = F_{\mathcal{B}}(\phi)$ and $cone(G(\phi)_{\bullet J'}) = F_{\mathcal{B}'}(\phi)$ for all $\phi \in \Phi'$. Thus, we also have that $k - p \le |J| \le k - p + 1$, $k - p \le |J'| \le k - p + 1$ and $dim(cone(G(\phi)_{\bullet J}) \cap cone(G(\phi)_{\bullet J'})) \ge k - p$. Furthermore, let $J^* = \begin{cases} J & \text{if } |J| \ge |J'| \\ J' & \text{otherwise} \end{cases}$ and recognize that we have either (i) $|J^*| = k - p$ and $span(G(\phi)_{\bullet J}) = span(G(\phi)_{\bullet J'})$ for all $\phi \in \Phi'$, or (ii) $|J^*| = k - p + 1$ and $span(cone(G(\phi)_{\bullet J^*})) = span(cone(G(\phi)_{\bullet (J \cup J')}))$ for all $\phi \in \Phi'$. More importantly, in either case we also have $aff\left(\bigcup_{\upsilon \in \Upsilon} q(\phi, \upsilon)\right) \subseteq span(G(\phi)_{\bullet J^*})$ for all $\phi \in \Phi'$. We now show that because $dim(\Phi') = p$, we actually have $aff\left(\bigcup_{\upsilon \in \Upsilon} q(\phi, \upsilon)\right) \subseteq span(G(\phi)_{\bullet J^*})$ for all $\phi \in \Phi$.

Recognize that because $aff\left(\bigcup_{\upsilon \in \Upsilon} q(\phi, \upsilon)\right) \subseteq span(G(\phi)_{\bullet J^*})$ for all $\phi \in \Phi'$, we have the following result.

For every $\upsilon \in \mathbb{R}^{k-p}$ and $\phi \in \Phi'$,

$$\exists\, \chi(\phi, \upsilon) \in \mathbb{R}^{|J^*|} \text{ such that } q(\phi, \upsilon) = G(\phi)_{\bullet J^*} \chi(\phi, \upsilon). \tag{6.3}$$

Since $dim(\Phi') = p$, we have that Φ' is full dimensional in $\mathbb{R}^p$. Hence, there must exist $\epsilon > 0$ and $\phi' \in \Phi'$ such that $B_\epsilon(\phi') \subseteq \Phi'$. Thus, for every $\phi \in \mathbb{R}^p$ we have $\phi' + \epsilon\phi \in \Phi'$. Thus, using (6.3) we obtain the following result.

$$q(\phi' + \epsilon\phi, \upsilon) = G(\phi' + \epsilon\phi)_{\bullet J^*}\chi(\phi' + \epsilon\phi, \upsilon) \text{ for all } \phi \in \mathbb{R}^p \text{ and } \upsilon \in \mathbb{R}^{k-p} \tag{6.4}$$

We now assume without loss of generality that $\phi' = \mathbf{0}$ and $\epsilon = 1$. If this were not the case, recognize that it could easily be achieved by constructing an equivalent mpLCP using a simple change of parameters in which we replace ϕ with $\frac{\phi-\phi'}{\epsilon}$. Under this assumption, (6.4) shows that for every $\phi \in \mathbb{R}^p$, each point in $\bigcup_{\upsilon\in\Upsilon} q(\phi, \upsilon)$ can be represented as a linear combination of the columns of $G(\phi)_{\bullet J^*}$. Hence, $aff\left(\bigcup_{\upsilon\in\Upsilon} q(\phi, \upsilon)\right) \subseteq span(G(\phi)_{\bullet J^*})$ for all $\phi \in \mathbb{R}^p \supset \Phi$. □

Proposition 6.2 *Let distinct complementary bases $\mathcal{B}$ and $\mathcal{B}'$ be given such that the invariancy regions $\mathcal{IR}_{\mathcal{B}}$ and $\mathcal{IR}_{\mathcal{B}'}$ overlap. Without loss of generality, there exists a partition of Θ that does not include $\mathcal{IR}_{\mathcal{B}'}$.*

Proof The claim of the proposition is trivial when $\mathcal{IR}_{\mathcal{B}'} \subseteq \mathcal{IR}_{\mathcal{B}}$ and so we assume this is not the case. From the arguments used in the proof of Proposition 6.1, we can conclude that for every $\theta = (\phi, \upsilon) \in \mathcal{IR}_{\mathcal{B}'}$, υ lies on a facet of $\mathcal{C}_{\mathcal{B}'}(\phi)$ whose dimension is either $k - p$ or $k - p + 1$. Thus, since $M(\phi)$ is sufficient for each $\phi \in \Phi$ and consequently $\mathcal{K}(M(\phi))$ is convex for each $\phi \in \Phi$, for every $\theta = (\phi, \upsilon) \in (\mathcal{IR}_{\mathcal{B}'} \setminus \mathcal{IR}_{\mathcal{B}})$ we must have that there exists a set $\mathcal{B}(\theta)$ of complementary bases such that: (i) $\upsilon \in \mathcal{C}_{\mathcal{B}^*}(\phi)$ for all $\mathcal{B}^* \in \mathcal{B}(\theta)$, and (ii) $|\mathcal{B}(\theta) \setminus \mathcal{B}'| \geq 1$. Furthermore, since there are a finite number of bases there must exist at least one basis $\tilde{\mathcal{B}} \in \mathcal{B}(\theta)$ such that $dim\left(\mathcal{C}_{\tilde{\mathcal{B}}}(\phi) \cap \bigcup_{\upsilon\in\Upsilon} q(\phi, \upsilon)\right) = k - p$ for all $\phi \in \Phi$ such that $dim\left(\mathcal{C}_{\mathcal{B}'}(\phi) \cap \bigcup_{\upsilon\in\Upsilon} q(\phi, \upsilon)\right) = k - p$. Since $\mathcal{IR}_{\mathcal{B}'}$ is full dimensional, this shows that $\mathcal{IR}_{\tilde{\mathcal{B}}}$ is also full dimensional. Hence, we use the following strategy for including invariancy regions in our partition of Θ: (i) Let $\mathscr{K} = \mathcal{IR}_{\mathcal{B}}$; (ii) Select a $\theta \in (\mathcal{IR}_{\mathcal{B}'} \setminus \mathscr{K})$; (iii) include $\mathcal{IR}_{\tilde{\mathcal{B}}}$ in the partition of Θ and add $\mathcal{IR}_{\tilde{\mathcal{B}}}$ to $\mathscr{K}$; (iv) If $(\mathcal{IR}_{\mathcal{B}'} \setminus \mathscr{K}) \neq \emptyset$, go back to Step (ii). Thus, by following this strategy we ensure that, although $\mathcal{IR}_{\mathcal{B}'}$ is not included in the partition of Θ, for every $\theta \in \mathcal{IR}_{\mathcal{B}'}$ there is a full dimensional invariancy region included in the partition of Θ that contains θ. □

Recognize that the result of Proposition 6.2 leads to the following corollary.

Corollary 6.1 *If an instance of mpLCP is feasible and $M(\phi)$ is a sufficient matrix for all $\phi \in \Phi$, then there exists a true partition of Θ.*

Proof Assume that a given instance of mpLCP is feasible and $M(\phi)$ is a sufficient matrix for all $\phi \in \Phi$. Also assume that a f.c.b. $\mathcal{B}$ exists such that $\mathcal{IR}_{\mathcal{B}}$ is full dimensional, because otherwise Θ contains only one feasible point and thus its partition is trivial. The procedures outlined in Chapters 4 and 5 guarantee that a partition of Θ exists and can be generated. Now, given any quasi-partition of Θ, utilize the following strategy to obtain a true partition: (i) select a pair of distinct f.c.b.'s $\mathcal{B}$ and $\mathcal{B}'$ such that the invariancy regions $\mathcal{IR}_{\mathcal{B}}$ and $\mathcal{IR}_{\mathcal{B}'}$ overlap, (ii) utilize the procedure outlined at the end of the proof of Proposition 6.2 to generate a new partition that does not include $\mathcal{IR}_{\mathcal{B}'}$, (iii) if the partition obtained in step (ii) is not a true partition, go back to step (i). □

The result of Corollary 6.1 provides an answer to Question 6.1. Note that we have not studied whether or not the converse of Corollary 6.1 holds. The result we have obtained is certainly satisfactory, though, since we already assume in this work that $M(\phi)$ is sufficient for all $\phi \in \Phi$.

We now move our focus to Question 6.2. It seems that the best strategy for ensuring that a true partition is discovered when attempting to partition Θ is this: Each time a full dimensional invariancy region is being considered during phase 2, discard all other invariancy regions that overlap the one currently being considered. One very naive way in which this could be accomplished is given in the following steps.

1. Given a f.c.b. $\mathcal{B}$ for which $dim(\mathcal{IR}_{\mathcal{B}}) = k$, let $\mathscr{B} = \{\mathcal{B}\}$ and create a modified mpLCP in which Θ is replaced by $\mathcal{IR}_{\mathcal{B}}$.
2. Perform phase 1 of our two phase mpLCP procedure. (Recall that $\mathscr{B}$ denotes the set of bases not considered in our procedure.)
3. If phase 1 returns an initial basis $\mathcal{B}'$ with $dim(\mathcal{IR}_{\mathcal{B}'}) = k$, then $\mathcal{IR}_{\mathcal{B}}$ and $\mathcal{IR}_{\mathcal{B}'}$ overlap. Add $\mathcal{B}'$ to $\mathscr{B}$ and repeat Step (2). Otherwise, if phase 1 does not return an initial basis, STOP.

This strategy then provides an answer to Question 6.2. We point out, though, that this may not be the most appropriate response to Question 6.2. It seems that a better response, although somewhat disappointing, should be that, even though one can ensure that a true partition can be obtained, it is generally impractical to do so. In the worst case the invariancy regions associated with every f.c.b. can overlap. To see this, recall Figure 3.1. In this figure, it could easily be the case that all parametric complementary cones, even those not depicted, could have a 1-dimensional intersection with $\bigcup_{\upsilon \in \Upsilon} q(\phi, \upsilon)$ for all $\phi \in \Phi$. Even though this result seems a bit disappointing, we do point out that obtaining a quasi-partition for mpLCP still provides a solution for every $\theta \in \Theta$ and is therefore perfectly acceptable. Before ending this section, we do provide a few key results that establish situations in which we can be sure that overlapping regions either do not exist or will not be obtained by steps in our algorithms. We then provide one final result that offers a practical step that can be taken in order to eliminate from consideration some, though not all, bases whose invariancy regions overlap a region that is currently under consideration.

Proposition 6.3 *For a complementary basis $\mathcal{B}$, if $Z_{\mathcal{B}} = \emptyset$ (recall* (4.4)*) then there does not exist another complementary basis $\mathcal{B}'$ such that $\mathcal{IR}_{\mathcal{B}}$ and $\mathcal{IR}_{\mathcal{B}'}$ overlap.*

Proof Recognize that the set $Z_{\mathcal{B}}$ can be interpreted as the set of $i \in \mathcal{B}$ such that for all $\phi \in \Phi$, $G(\phi)_{\bullet i}$ is unnecessary in the representation of the points in $\bigcup_{\upsilon \in \Upsilon} q(\phi, \upsilon)$ as linear combinations of the columns of $G(\phi)_{\bullet \mathcal{B}}$, i.e., $\bigcup_{\upsilon \in \Upsilon} q(\phi, \upsilon) \subseteq span\left(G(\phi)_{\bullet(\mathcal{B}\setminus\{i\})}\right)$ for all $\phi \in \Phi$. Thus, if $Z_{\mathcal{B}} = \emptyset$ then $\bigcup_{\upsilon \in \Upsilon} q(\phi, \upsilon)$ intersects the relative interior of $\mathcal{C}_{\mathcal{B}}(\phi)$ for all but at most a finite number of $\phi \in \Phi$. Hence, there cannot exist $\Phi' \subseteq \Phi$ such that $dim(\Phi') = p$ and $dim\left(\bigcup_{\upsilon \in \Upsilon} q(\phi, \upsilon) \cap \mathcal{C}_{\mathcal{B}}(\phi) \cap \mathcal{C}_{\mathcal{B}'}(\phi)\right) = k - p$ for all $\phi \in \Phi'$. Thus, the result of the proposition follows from Lemma 6.1. □

Proposition 6.4 *Given a f.c.b. $\mathcal{B}$ for which $dim(\mathcal{IR}_{\mathcal{B}}) = k$ and an $i \in \mathcal{B} \setminus Z_{\mathcal{B}}$, neither a diagonal pivot involving i, as outlined in condition (1) of Proposition 4.7, nor an exchange pivot involving i, as outlined in condition (2) of Proposition 4.7, will result in a new basis $\mathcal{B}'$ for which $\mathcal{IR}_{\mathcal{B}}$ and $\mathcal{IR}_{\mathcal{B}'}$ overlap.*

Proof Since $i \notin Z_{\mathcal{B}}$ there are at most a finite number of $\phi \in \Phi$ such that $G(\phi)_{\bullet i}$ is unnecessary in the representation of the points in $\bigcup_{\upsilon \in \Upsilon} q(\phi, \upsilon)$ as linear combinations of the columns of $G(\phi)_{\bullet \mathcal{B}}$. This shows that there are at most a finite number of $\phi \in \Phi$ such that $dim\left(\bigcup_{\upsilon \in \Upsilon} q(\phi, \upsilon) \cap cone\left(G(\phi)_{\bullet(\mathcal{B}\setminus\{i\})}\right)\right) = k - p$, i.e., there is no $\Phi' \subseteq \Phi$ such that $dim(\Phi') = p$ and $dim\left(\bigcup_{\upsilon \in \Upsilon} q(\phi, \upsilon) \cap cone\left(G(\phi)_{\bullet(\mathcal{B}\setminus\{i\})}\right)\right) = k - p$ for all $\phi \in \Phi'$. Recall that if $\mathcal{B}'$ is obtained from $\mathcal{B}$ by a diagonal pivot involving i, as outlined in condition (1) of Proposition 4.7, or an exchange pivot involving i, as outlined in condition (2) of Proposition 4.7, then $(\mathcal{C}_{\mathcal{B}}(\phi) \cap \mathcal{C}_{\mathcal{B}'}(\phi)) \subseteq cone\left(G(\phi)_{\bullet(\mathcal{B}\setminus\{i\})}\right)$ for all $\phi \in \Phi$. Thus, there cannot exist a $\Phi' \subseteq \Phi$ such that $dim(\Phi') = p$ and $dim\left(\bigcup_{\upsilon \in \Upsilon} q(\phi, \upsilon) \cap \mathcal{C}_{\mathcal{B}}(\phi) \cap \mathcal{C}_{\mathcal{B}'}(\phi)\right) = k - p$ for all $\phi \in \Phi'$. The result of the proposition then follows from Lemma 6.1. □

Proposition 6.5 *Given a f.c.b. $\mathcal{B}$ for which $dim(\mathcal{IR}_{\mathcal{B}}) = k$, if there exists an index $i \in Z_{\mathcal{B}}$ such that $(T_{\mathcal{B}}(\phi))_{i,\bar{i}}$ is not identically zero, then for the basis $\mathcal{B}' = (\mathcal{B} \setminus \{i\}) \cup \{\bar{i}\}$ we have that $\mathcal{IR}_{\mathcal{B}}$ and $\mathcal{IR}_{\mathcal{B}'}$ overlap.*

Proof Since $i \in Z_{\mathcal{B}}$ we have $\bigcup_{\upsilon \in \Upsilon} q(\phi, \upsilon) \subseteq span\left(G(\phi)_{\bullet(\mathcal{B}\setminus\{i\})}\right)$ for all $\phi \in \Phi$. Furthermore, since $dim(\mathcal{IR}_{\mathcal{B}}) = k$ there must exist $\Phi' \subset \Phi$ such that $dim(\Phi') = p$ and $dim\left(\bigcup_{\upsilon \in \Upsilon} q(\phi, \upsilon) \cap cone(G(\phi)_{\bullet(\mathcal{B}\setminus\{i\})})\right) = k - p$ for all $\phi \in \Phi'$. Since

$(T_{\mathcal{B}}(\phi))_{i,\bar{\imath}}$ is not identically zero, $\mathcal{B}' = (\mathcal{B}\setminus\{i\})\cup\{\bar{\imath}\}$ is a basis and $\mathcal{C}_{\mathcal{B}}(\phi)\cap\mathcal{C}_{\mathcal{B}'}(\phi) = cone\left(G(\phi)_{\bullet(\mathcal{B}\setminus\{i\})}\right)$ for all $\phi \in \Phi$. Thus, $dim\left(\bigcup_{\upsilon\in\Upsilon} q(\phi,\upsilon)\cap\mathcal{C}_{\mathcal{B}}(\phi)\cap\mathcal{C}_{\mathcal{B}'}(\phi)\right) = k-p$ for all $\phi \in \Phi'$. The result of the proposition then follows from Lemma 6.1. □

As a result of Proposition 6.5, the procedures we presented in Chapters 4 and 5 can be modified so that when a basis $\mathcal{B}$ is discovered for which there exists an index $i \in Z_{\mathcal{B}}$ such that $(T_{\mathcal{B}}(\phi))_{i,\bar{\imath}}$ is not identically zero, the basis $\mathcal{B}' = (\mathcal{B} \setminus \{i\}) \cup \{\bar{\imath}\}$ is added to $\mathscr{B}$ and is not considered for inclusion in the final partition of Θ. The implementation we discuss in the next chapter incorporates this modification. We also point out that this modification is enough to ensure that the overlapping regions we discovered during our phase 1 consideration of Example 2.1 are not both considered. Observe from Table A.1 in Appendix A that $w_3 \in Z^{ph1}_{\mathcal{B}^{*2.1}}$ and $(T^{ph1}_{\mathcal{B}^{*2.1}}(\phi))_{w_3,z_3}$ is not identically zero. Hence, we add $(\mathcal{B}^{*2.1} \setminus \{w_3\}) \cup \{z_3\} = \mathcal{B}^{2.1}_{ii}$ to $\mathscr{B}$ during our initial consideration of $\mathcal{B}^{*2.1}$ and as a result, the invariancy region associated with this basis would not be considered throughout the remainder of phase 1.

Chapter 7
Assessment of Performance

In this chapter we discuss the performance of the aforementioned procedures. We begin with empirical results and conclude with a theoretical discussion of computational complexity.

7.1 Experimental Results

We now present the results of a computational experiment we conducted in order to test the practical performance of the proposed algorithms. We also include a few brief notes on our implementation.

We implemented the proposed two-phase algorithm using MATLAB. MATLAB code for this implementation has been made freely available at https://github.com/Nadelgren/mpLCP_solver. Within this implementation, all auxiliary NLPs are solved using the "fmincon" function within MATLAB. We note that since the version of mpLCP we consider in this work was previously unsolved, there is no other method with which we can compare. All tests were run using MATLAB R2016a [41] on a machine running Linux Mint 17 with two 2.4 GHz processors, each with 4 GB of RAM.

For our experiment, we randomly generated 105 instances. We produced ten instances for each value of h in $\{4, \dots, 12\}$: half with $k = 2$ and half with $k = 3$. We also produced an additional five instances for each value of h in $\{13, 14, 15\}$ with $k = 2$. Each instance was derived from a multiobjective program with $k + 1$ convex quadratic objectives. These multiobjective programs were then scalarized using the weighted-sum method (see, for example, [17]) to obtain an mpQP in the form of (1.5) and then reformulated as an instance of mpLCP. We then solved each instance using our implementation of the proposed method. A summary of the results is given in Table 7.1. As expected, the results display a positive correlation between the instance size and (i) the numbers of iterations, (ii) the number of invariancy

N. Adelgren, *Advancing Parametric Optimization*, SpringerBriefs in Optimization,
https://doi.org/10.1007/978-3-030-61821-6_7

Table 7.1 Experimental Results—Averages are taken over instances that were solved in less than one hour

k	h	Number solved in < 1 hour	Average time (s)	Average num. Ph1 iterations	Average num. Ph2 iterations	Average num. regions	Average time per iteration (s)
2	4	5	7.1	1.2	3.0	1.8	1.6
	5	5	25.5	1.8	3.0	3.0	5.3
	6	5	50.7	2.0	6.6	4.8	5.8
	7	5	144.0	1.8	7.2	6.4	16.0
	8	5	66.5	2.0	4.2	4.2	10.7
	9	5	215.2	3.0	10.6	8.8	15.8
	10	5	132.4	2.0	6.8	5.8	15.0
	11	5	347.2	4.6	9.6	7.4	24.4
	12	4	460.8	2.2	12.2	10.0	31.7
	13	5	1,329.8	4.2	18.2	17.4	59.3
	14	4	816.5	3.0	13.7	12.2	48.0
	15	4	1,547.2	4.0	14.7	14.0	82.5
3	4	5	21.4	1.2	5.0	4.0	3.4
	5	5	47.7	3.6	5.0	4.4	5.5
	6	5	51.6	1.4	4.2	3.4	9.2
	7	5	212.0	1.6	8.6	8.0	20.7
	8	5	417.8	3.2	8.6	7.4	35.4
	9	5	1,406.8	2.8	15.8	12.0	75.6
	10	5	887.9	1.8	10.4	9.8	80.4
	11	4	2,124.4	2.0	14.2	11.7	130.7
	12	2	1,648.2	1.5	8.5	8.0	164.8

regions, and (iii) the average CPU time spent in each iteration. Figure 7.1 depicts the partitions of Θ computed during this experiment for four instances. Recall that for each pair of k and h values described in Table 7.1 we generated five instances of mpLCP. The label on each subfigure of Figure 7.1 indicates which of the five instances has its solution depicted in the figure.

We now discuss a few details of our implementation. All parameters of the "fmincon" optimization function were left at their default values, except the constraint violation tolerance, which was set to 10^{-9}; the maximum number of iterations, which was set to 4,000; and the maximum number of function evaluations, which was set to 8,000. Also, when solving NLPs as feasibility problems, we assumed that λ was sufficiently large when it reached a value of 10^{-4}. Additionally, in our implementation, we explicitly compute the tableau associated with each discovered basis. Thus, the overall performance could likely be improved by instead using matrix factorization techniques. In the computational experiments conducted in [2], we discovered that the efficiency of phase 1 could be improved if we sorted k-dimensional boundaries of $(k + 1)$-dimensional regions based on the ρ component of the normalized normal vector of the boundaries (recall that in [2] all boundaries

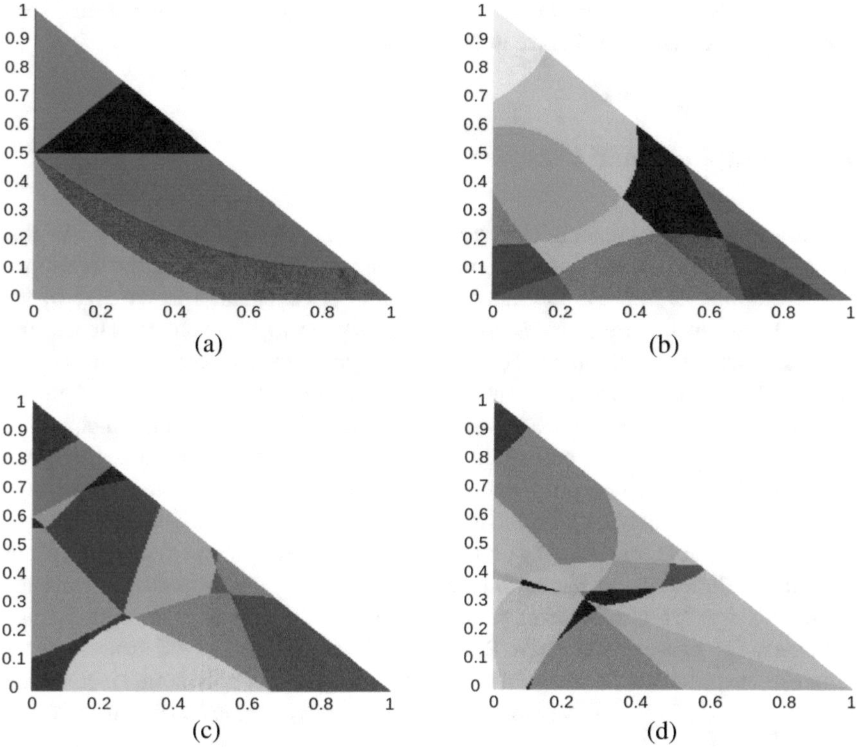

Fig. 7.1 Partitions of Θ. (**a**) $k = 2, h = 6$: Instance 3 of 5. (**b**) $k = 2, h = 9$: Instance 5 of 5. (**c**) $k = 2, h = 11$: Instance 4 of 5. (**d**) $k = 2, h = 13$: Instance 1 of 5

are hyperplanes) prior to seeking adjacent regions across each boundary. We used a similar strategy here, but rather than sorting using the ρ component of the normal vector, we use the ρ component of the normalized gradient vector of each polynomial function defining a k-dimensional boundary of a given region, evaluated at the solution of NLP_s. By searching for adjacent regions across k-dimensional boundaries with the largest ρ component of this normalized gradient first, we increase the likelihood of discovering a new region that has a lower optimal value of NLP_s (minimal value of ρ) than the previous region. In this way, we aim to speed up the process of discovering a region that intersects the hyperplane $\rho = 0$.

We also note that in our implementation, depending on the optimization technique chosen for use alongside MATLAB's "fmincon" function (interior point, trust region reflective, SQP, active set), the optimizer occasionally returns a local optimal solution that is not always feasible, even in situations in which a feasible solution exists. We found that this occurred more frequently for larger problems. To counter this issue, each time we use the "fmincon" function, we begin with the SQP optimizer, but if the function fails or returns an infeasible solution, we resolve using the interior point optimizer. This is, of course, inefficient, and we will therefore seek a more robust optimizer to utilize in future implementations.

7.2 Computational Complexity

We now discuss the theoretical performance of the algorithms presented in this work. We begin by noting that a polynomial time algorithm can never be developed for partitioning Θ because the number of bases is exponential in the size of the problem, and in the worst case, all bases may need to be explored. Hence, our discussion on complexity focuses on the work required to *process* a given invariancy region. Where, by *process*, we mean the computational effort required to: (i) determine the dimension of the region, (ii) compute the set of defining inequalities for the region that should be exploited when searching for adjacent regions, and (iii) for each defining inequality identified in (ii), determine an additional invariancy region (if one exists) that is adjacent to the first along the hypersurface defined by the inequality. We note that the set of inequalities identified in item (ii) above depends on both the dimension of the region and which phase of the procedure is currently being considered. For this reason, we consider the complexity of processing a region separately for phase 1 and phase 2, beginning with phase 2. Furthermore, in the propositions presented below, we assume that the number of polynomial inequalities used to define Θ is n, and we employ the notation $T_{NLP}(var, con)$ (similar to that of [2, 11]) to denote the time required to solve a nonlinear program in var variables, with a linear objective function and con polynomial constraints.

Proposition 7.1 *In the worst case, the computational effort required to process a k-dimensional invariancy region during phase 2 is*

$$(h^2 + 1)T_{NLP}(k + 1, h + n) + h(h - 1)T_{NLP}(k + 1, 2h + n - 1), \tag{7.1}$$

and the computational effort required to process a $(k - 1)$-dimensional invariancy region during phase 2 is

$$T_{NLP}(k+1, h+n)+(h-1)T_{NLP}(k+1, h+n+1)+(h-1)^2 T_{NLP}(k+1, 2h+n-1). \tag{7.2}$$

Proof We begin by noting that in both the k- and $(k - 1)$-dimensional cases, the dimension of the invariancy region is determined by solving NLP_D (4.19), thus accounting for $T_{NLP}(k + 1, h + n)$ time in each case. Let us now focus on the k-dimensional case. Recognize from Algorithms 1, 2, and 3 that the additional h^2 occurrences of $T_{NLP}(k + 1, h + n)$ in (7.1) come from: (i) solving NLP_F (4.7) up to h times (line 2 of Algorithm 2) and (ii) solving NLP_H (4.12) up to $h(h - 1)$ times (line 5 of Algorithm 3). Furthermore, recognize from Algorithms 1 and 4 that the $h(h - 1)$ occurrences of $T_{NLP}(k + 1, 2h + n - 1)$ in (7.1) come from: (i) up to

h calls to GETADJACENTREGIONSACROSS (line 5 of Algorithm 1) and (ii) solving NLP_A (4.14) up to $h-1$ times (line 9 of Algorithm 4).

We now consider the $(k-1)$-dimensional case. Recall that in this case a defining inequality of the invariancy region is known whose associated hypersurface forms a superset of the invariancy region itself. Hence, it is not necessary to construct the set of non-redundant inequalities for the invariancy region. Instead NLP_G (4.23) is used to determine additional defining inequalities whose associated hypersurfaces have $(k-1)$-dimensional intersection with both the hypersurface associated with the aforementioned defining inequality and the invariancy region itself. Thus, recognize from Algorithm 6 that: (i) the $h-1$ occurrences of $T_{NLP}(k+1, h+n+1)$ in (7.2) come from solving NLP_G (4.23) up to $h-1$ times (line 3 of Algorithm 6) and (ii) the $(h-1)^2$ occurrences of $T_{NLP}(k+1, 2h+n-1)$ in (7.2) come from solving NLP_{A2} (4.24) up to $(h-1)^2$ times (line 11 of Algorithm 6). □

Proposition 7.2 *In the worst case, the computational effort required to process a $(k+1)$-dimensional invariancy region during phase 1 is*

$$T_{NLP}(k+1, h+n)+(h^2+1)T_{NLP}(k+2, h+n)+h(h-1)T_{NLP}(k+2, 2h+n-1), \tag{7.3}$$

and the computational effort required to process a k-dimensional invariancy region during phase 1 is

$$\begin{aligned} &T_{NLP}(k+1, h+n) + T_{NLP}(k+2, h+n)+ \\ &(h-1)T_{NLP}(k+2, h+n+1) + (h-1)^2 T_{NLP}(k+2, 2h+n-1). \end{aligned} \tag{7.4}$$

Proof The results presented in this proposition follow directly from the results of Proposition 7.1 and the following two facts: (i) in phase 1, an extra nonlinear program, NLP_S (5.4), is solved to identify a subset of invariancy regions' defining inequalities that are most likely to yield adjacent phase 1 invariancy regions whose intersection with the hyperplane $\rho = 0$ is k-dimensional and (ii) in all other ways, phase 1 is a special case of phase 2 in which $k+1$ parameters are considered rather than k. □

Chapter 8
Conclusion

In this work, we have introduced the first ever method for solving mpLCP (1.1) in which all elements of the matrix M and the vector q are permitted to be affine functions of the parameters, so long as $M(\theta)$ is a sufficient matrix for each permissible value of θ. Phase 1 answers the previously unanswered question of how one can determine an initial full dimensional invariancy region that can be used as a starting point in the process of partitioning the parameter space Θ. The partition of Θ is carried out in phase 2.

We also discuss a few difficulties that arise during the execution of phases 1 and 2 and provide some initial insight into what causes these issues and how they can either be avoided or dealt with. We note that some of these issues do leave open research questions that we intend to study in future work. Namely, we ask whether or not there are restrictions that can be placed on the structure of $M(\theta)$, $q(\theta)$, and/or Θ to ensure that either $\hat{\Theta}$, the feasible subset of Θ, is nonempty and connected or no overlapping invariancy regions exist. Additionally, experimental and theoretical results are provided that, respectively, give evidence of the utility of the proposed method in practice and provide worst-case complexity analyses.

N. Adelgren, *Advancing Parametric Optimization*, SpringerBriefs in Optimization,
https://doi.org/10.1007/978-3-030-61821-6_8

Appendix A
Tableaux for Example 2.1

N. Adelgren, *Advancing Parametric Optimization*, SpringerBriefs in Optimization,
https://doi.org/10.1007/978-3-030-61821-6

Table A.1 $T^{ph1}_{\mathcal{B}^{*2.1}}(\phi, \rho)$

	w_1	w_2	w_3	w_4	w_5	z_1	z_2	z_3	z_4	z_5	RHS
w_1	1	0	0	0	0	0	0	-1	-3	5	3
w_2	0	1	0	0	0	0	0	-2	-2	-2	$-\phi_1 - 2 + 3\rho$
w_3	0	0	1	0	0	1	2	$\phi_2 - 2\phi_1 - 4$	$2\phi_2 - \phi_1 - 3$	$2 - 4\phi_2 - 3\phi_1$	0
w_4	0	0	0	1	0	3	2	$2\phi_2 - \phi_1 - 3$	$\phi_1 - \phi_2 - 4$	$3 - 4\phi_2 - 3\phi_1$	0
w_5	0	0	0	0	1	-5	2	$2 - 4\phi_2 - 3\phi_1$	$3 - 4\phi_2 - 3\phi_1$	$\phi_2 - 3$	0

Table A.2 $T^{ph1}_{\mathcal{B}_i^{2.1}}(\phi, \rho)$

	w_1	w_2	w_3	w_4	w_5	z_1	z_2	z_3	z_4	z_5	RHS
w_1	1	$-\frac{1}{2}$	0	0	0	0	0	0	-2	6	$\frac{\phi_1}{2} - \frac{3\rho}{2} + 4$
z_2	0	$\frac{\phi_2}{4} - \frac{\phi_1}{2} - 1$	$\frac{1}{2}$	0	0	$\frac{1}{2}$	1	0	$\frac{\phi_1}{2} + \frac{\phi_2}{2} + \frac{1}{2}$	$3 - \frac{5\phi_2}{2} - \frac{\phi_1}{2}$	$\frac{(2\phi_1 - \phi_2 + 4)\left(\frac{\phi_1}{2} - \frac{3\rho}{2} + 1\right)}{2}$
z_3	0	$-\frac{1}{2}$	0	0	0	0	0	1	1	1	$\frac{\phi_1}{2} - \frac{3\rho}{2} + 1$
w_4	0	$\frac{\phi_1}{2} + \frac{\phi_2}{2} + \frac{1}{2}$	-1	1	0	2	0	0	$\phi_1 - 4\phi_2 - 2$	$-\phi_1 - \phi_2$	$-\frac{(\phi_1 + \phi_2 + 1)(\phi_1 - 3\rho + 2)}{2}$
w_5	0	$3 - \frac{5\phi_2}{2} - \frac{\phi_1}{2}$	-1	0	1	-6	0	0	$-\phi_1 - \phi_2$	$4\phi_1 + 10\phi_2 - 11$	$\frac{(\phi_1 + 5\phi_2 - 6)(\phi_1 - 3\rho + 2)}{2}$

Table A.3 $T^{ph1}_{\mathcal{B}^{2.1}_{ii}}(\phi, \rho)$ (Due to space limitations, we only display columns associated with nonbasic variables.)

	w_3	z_1	z_2	z_4	z_5	RHS
w_1	$-\frac{1}{2\phi_1-\phi_2+4}$	$-\frac{1}{2\phi_1-\phi_2+4}$	$-\frac{2}{2\phi_1-\phi_2+4}$	$-\frac{5\phi_1-\phi_2+9}{2\phi_1-\phi_2+4}$	$\frac{13\phi_1-\phi_2+18}{2\phi_1-\phi_2+4}$	3
w_2	$-\frac{2}{2\phi_1-\phi_2+4}$	$-\frac{2}{2\phi_1-\phi_2+4}$	$-\frac{4}{2\phi_1-\phi_2+4}$	$-\frac{2\phi_1+2\phi_2+2}{2\phi_1-\phi_2+4}$	$\frac{2\phi_1+10\phi_2-12}{2\phi_1-\phi_2+4}$	$3\rho-\phi_1-2$
z_3	$-\frac{1}{2\phi_1-\phi_2+4}$	$-\frac{1}{2\phi_1-\phi_2+4}$	$-\frac{2}{2\phi_1-\phi_2+4}$	$\frac{\phi_1-2\phi_2+3}{2\phi_1-\phi_2+4}$	$\frac{3\phi_1+4\phi_2-2}{2\phi_1-\phi_2+4}$	0
w_4	$-\frac{\phi_1-2\phi_2+3}{2\phi_1-\phi_2+4}$	$\frac{5\phi_1-\phi_2+9}{2\phi_1-\phi_2+4}$	$\frac{2\phi_1+2\phi_2+2}{2\phi_1-\phi_2+4}$	$\frac{3\phi_1^2-7\phi_1\phi_2+2\phi_1+5\phi_2^2-12\phi_2-7}{2\phi_1-\phi_2+4}$	$-\frac{3\phi_1^2+7\phi_1\phi_2-\phi_1+4\phi_2^2+3\phi_2-6}{2\phi_1-\phi_2+4}$	0
w_5	$-\frac{3\phi_1+4\phi_2-2}{2\phi_1-\phi_2+4}$	$-\frac{13\phi_1-\phi_2+18}{2\phi_1-\phi_2+4}$	$-\frac{2\phi_1+10\phi_2-12}{2\phi_1-\phi_2+4}$	$-\frac{3\phi_1^2+7\phi_1\phi_2-\phi_1+4\phi_2^2+3\phi_2-6}{2\phi_1-\phi_2+4}$	$-\frac{-9\phi_1^2-26\phi_1\phi_2+18\phi_1-15\phi_2^2+9\phi_2+8}{2\phi_1-\phi_2+4}$	0

Table A.4 $T^{ph1}_{\mathcal{B}^{2.1}_{iii}}(\phi, \rho)$. (Due to space limitations, we only display columns associated with nonbasic variables.)

	w_2	w_3	w_4	z_1	z_5	RHS
w_1	$-\frac{\phi_1+6\phi_2+4}{2(4\phi_2-\phi_1+2)}$	$\frac{2}{4\phi_2-\phi_1+2}$	$-\frac{2}{4\phi_2-\phi_1+2}$	$-\frac{4}{4\phi_2-\phi_1+2}$	$\frac{2(\phi_1+\phi_2)}{4\phi_2-\phi_1+2}+6$	$\frac{\phi_1}{2}-\frac{3\rho}{2}+\frac{(\phi_1+\phi_2+1)(\phi_1-3\rho+2)}{4\phi_2-\phi_1+2}+4$
z_2	$\frac{\phi_2}{4}-\frac{\phi_1}{2}+\frac{\left(\frac{\phi_1}{2}+\frac{\phi_2}{2}+\frac{1}{2}\right)^2}{4\phi_2-\phi_1+2}-1$	$\frac{3\phi_2-2\phi_1+1}{2(4\phi_2-\phi_1+2)}$	$\frac{\phi_1+\phi_2+1}{8\phi_2-2\phi_1+4}$	$\frac{\phi_1+6\phi_2+4}{2(4\phi_2-\phi_1+2)}$	$-\frac{9\phi_1-13\phi_2+\phi_1\phi_2+21\phi_2^2-12}{2(4\phi_2-\phi_1+2)}$	$-\frac{(\phi_1-3\rho+2)(3\phi_1^2-7\phi_1\phi_2+2\phi_1+5\phi_2^2-12\phi_2-7)}{4(4\phi_2-\phi_1+2)}$
z_3	$-\frac{3\phi_2-2\phi_1+1}{2(4\phi_2-\phi_1+2)}$	$-\frac{1}{4\phi_2-\phi_1+2}$	$\frac{1}{4\phi_2-\phi_1+2}$	$\frac{2}{4\phi_2-\phi_1+2}$	$1-\frac{\phi_1+\phi_2}{4\phi_2-\phi_1+2}$	$\frac{(3\phi_2-2\phi_1+1)(\phi_1-3\rho+2)}{2(4\phi_2-\phi_1+2)}$
z_4	$-\frac{\phi_1+\phi_2+1}{8\phi_2-2\phi_1+4}$	$\frac{1}{4\phi_2-\phi_1+2}$	$-\frac{1}{4\phi_2-\phi_1+2}$	$-\frac{2}{4\phi_2-\phi_1+2}$	$\frac{\phi_1+\phi_2}{4\phi_2-\phi_1+2}$	$\frac{(\phi_1+\phi_2+1)(\phi_1-3\rho+2)}{2(4\phi_2-\phi_1+2)}$
w_5	$-\frac{9\phi_1-13\phi_2+\phi_1\phi_2+21\phi_2^2-12}{2(4\phi_2-\phi_1+2)}$	$\frac{\phi_1+\phi_2}{4\phi_2-\phi_1+2}-1$	$-\frac{\phi_1+\phi_2}{4\phi_2-\phi_1+2}$	$-\frac{2(\phi_1+\phi_2)}{4\phi_2-\phi_1+2}-6$	$4\phi_1+10\phi_2+\frac{(\phi_1+\phi_2)^2}{4\phi_2-\phi_1+2}-11$	$\frac{(\phi_1-3\rho+2)(9\phi_1-13\phi_2+\phi_1\phi_2+21\phi_2^2-12)}{2(4\phi_2-\phi_1+2)}$

Table A.5 $T^{ph1}_{\mathcal{B}^{2.1}_{iv}}(\phi, \rho)$

	w_1	w_2	w_3	w_4	w_5	z_1	z_2	z_3	z_4	z_5	RHS
w_1	1	$-\frac{3}{2}$	0	0	0	0	0	2	0	8	$\frac{3\,\phi_1}{2} - \frac{9\,\rho}{2} + 6$
z_2	0	$\frac{\phi_1}{4} - \frac{\phi_2}{4} - 1$	0	$\frac{1}{2}$	0	$\frac{3}{2}$	1	$\frac{3\,\phi_2}{2} - \phi_1 + \frac{1}{2}$	0	$\frac{7}{2} - \frac{3\,\phi_2}{2} - 2\,\phi_1$	$\frac{\left(\frac{\phi_1}{2} - \frac{3\,\rho}{2} + 1\right)(\phi_2 - \phi_1 + 4)}{2}$
w_3	0	$\frac{3\,\phi_2}{2} - \phi_1 + \frac{1}{2}$	1	-1	0	-2	0	$\phi_1 - 4\,\phi_2 - 2$	0	$2\,\phi_1 - 3\,\phi_2 - 2$	$-\frac{(3\,\phi_2 - 2\,\phi_1 + 1)(\phi_1 - 3\,\rho + 2)}{2}$
z_4	0	$-\frac{1}{2}$	0	0	0	0	0	1	1	1	$\frac{\phi_1}{2} - \frac{3\,\rho}{2} + 1$
w_5	0	$\frac{7}{2} - \frac{3\,\phi_2}{2} - 2\,\phi_1$	0	-1	1	-8	0	$2\,\phi_1 - 3\,\phi_2 - 2$	0	$7\,\phi_1 + 8\,\phi_2 - 13$	$\frac{(4\,\phi_1 + 3\,\phi_2 - 7)(\phi_1 - 3\,\rho + 2)}{2}$

Table A.6 $T^{ph1}_{\mathcal{B}^{2.1}_v}(\phi)$. Note that this tableau also serves as a tableau during phase 2 of Example 2.1 ($T_{\mathcal{B}^{2.1}_2}(\phi)$), by substituting $\rho = 0$. (Due to space limitations, we only display columns associated with nonbasic variables.)

	w_2	w_4	w_5	z_1	z_3
w_1	$\frac{11\phi_1-17}{2(7\phi_1+8\phi_2-13)}$	$\frac{8}{7\phi_1+8\phi_2-13}$	$-\frac{8}{7\phi_1+8\phi_2-13}$	$\frac{64}{7\phi_1+8\phi_2-13}$	$-\frac{2(\phi_1-20\phi_2+5)}{7\phi_1+8\phi_2-13}$
z_2	$\frac{-9\phi_1^2-23\phi_1\phi_2+15\phi_1-17\phi_2^2+23\phi_2+3}{4(7\phi_1+8\phi_2-13)}$	$\frac{3\phi_1+5\phi_2-6}{2(7\phi_1+8\phi_2-13)}$	$\frac{4\phi_1+3\phi_2-7}{2(7\phi_1+8\phi_2-13)}$	$-\frac{11\phi_1-17}{2(7\phi_1+8\phi_2-13)}$	$\frac{-6\phi_1^2-\phi_1\phi_2+11\phi_1+15\phi_2^2-16\phi_2+1}{2(7\phi_1+8\phi_2-13)}$
w_3	$\frac{-6\phi_1^2-\phi_1\phi_2+11\phi_1+15\phi_2^2-16\phi_2+1}{2(7\phi_1+8\phi_2-13)}$	$-\frac{5\phi_1+11\phi_2-11}{7\phi_1+8\phi_2-13}$	$\frac{3\phi_2-2\phi_1+2}{7\phi_1+8\phi_2-13}$	$\frac{2(\phi_1-20\phi_2+5)}{7\phi_1+8\phi_2-13}$	$-\frac{-3\phi_1^2+8\phi_1\phi_2+19\phi_1+41\phi_2^2-24\phi_2-22}{7\phi_1+8\phi_2-13}$
z_4	$-\frac{3\phi_1+5\phi_2-6}{2(7\phi_1+8\phi_2-13)}$	$\frac{1}{7\phi_1+8\phi_2-13}$	$-\frac{1}{7\phi_1+8\phi_2-13}$	$\frac{8}{7\phi_1+8\phi_2-13}$	$\frac{5\phi_1+11\phi_2-11}{7\phi_1+8\phi_2-13}$
z_5	$-\frac{4\phi_1+3\phi_2-7}{2(7\phi_1+8\phi_2-13)}$	$-\frac{1}{7\phi_1+8\phi_2-13}$	$\frac{1}{7\phi_1+8\phi_2-13}$	$-\frac{8}{7\phi_1+8\phi_2-13}$	$-\frac{3\phi_2-2\phi_1+2}{7\phi_1+8\phi_2-13}$

	RHS
w_1	$\frac{-11\phi_1^2+37\phi_1+48\phi_2+33\phi_1\rho-51\rho-44}{2(7\phi_1+8\phi_2-13)}$
z_2	$\frac{23\phi_1\phi_2-46\phi_2-17\rho-33\phi_1+59\phi_1\rho+85\phi_2\rho+17\phi_1\phi_2^2+23\phi_1^2\phi_2-27\phi_1^2\rho-51\phi_2^2\rho+3\phi_1^2+9\phi_1^3+34\phi_2^2-69\phi_1\phi_2\rho-6}{4(7\phi_1+8\phi_2-13)}$
w_3	$-\frac{(\phi_1-3\rho+2)\left(-6\phi_1^2-\phi_1\phi_2+11\phi_1+15\phi_2^2-16\phi_2+1\right)}{2(7\phi_1+8\phi_2-13)}$
z_4	$\frac{(\phi_1-3\rho+2)(3\phi_1+5\phi_2-6)}{2(7\phi_1+8\phi_2-13)}$
z_5	$\frac{(\phi_1-3\rho+2)(4\phi_1+3\phi_2-7)}{2(7\phi_1+8\phi_2-13)}$

Table A.7 $T^{ph1}_{\mathcal{B}^{2.1}_{vi}}(\phi,\rho)$. Note that this tableau also serves as the first tableau for phase 2 of Example 2.1 ($T_{\mathcal{B}^{2.1}_0}(\phi)$), by substituting $\rho = 0$. (Due to space limitations, we only display columns associated with nonbasic variables.)

	w_2	w_3	w_4	w_5
w_1	$-\frac{3\phi_1^2+18\phi_1\phi_2-37\phi_1-75\phi_2^2+64\phi_2+28}{2(-3\phi_1^2+8\phi_1\phi_2+19\phi_1+41\phi_2^2-24\phi_2-22)}$	$-\frac{2(\phi_1-20\phi_2+5)}{-3\phi_1^2+8\phi_1\phi_2+19\phi_1+41\phi_2^2-24\phi_2-22}$	$-\frac{2\phi_1+14\phi_2-22}{-3\phi_1^2+8\phi_1\phi_2+19\phi_1+41\phi_2^2-24\phi_2-22}$	$-\frac{2(13\phi_2-2\phi_1+6)}{-3\phi_1^2+8\phi_1\phi_2+19\phi_1+41\phi_2^2-24\phi_2-22}$
z_2	$\frac{9\phi_1^3-9\phi_1^2\phi_2-33\phi_1^2-87\phi_1\phi_2^2+21\phi_1\phi_2+22\phi_1-59\phi_2^3+13\phi_2^2+50\phi_2+5}{4(-3\phi_1^2+8\phi_1\phi_2+19\phi_1+41\phi_2^2-24\phi_2-22)}$	$\frac{-6\phi_1^2-\phi_1\phi_2+11\phi_1+15\phi_2^2-16\phi_2+1}{2(-3\phi_1^2+8\phi_1\phi_2+19\phi_1+41\phi_2^2-24\phi_2-22)}$	$\frac{3\phi_1^2+8\phi_1\phi_2-\phi_1+5\phi_2^2+5\phi_2-11}{-6\phi_1^2+16\phi_1\phi_2+38\phi_1+82\phi_2^2-48\phi_2-44}$	$\frac{9\phi_1-13\phi_2+\phi_1\phi_2+21\phi_2^2-12}{2(-3\phi_1^2+8\phi_1\phi_2+19\phi_1+41\phi_2^2-24\phi_2-22)}$
z_3	$-\frac{-6\phi_1^2-\phi_1\phi_2+11\phi_1+15\phi_2^2-16\phi_2+1}{2(-3\phi_1^2+8\phi_1\phi_2+19\phi_1+41\phi_2^2-24\phi_2-22)}$	$-\frac{7\phi_1+8\phi_2-13}{-3\phi_1^2+8\phi_1\phi_2+19\phi_1+41\phi_2^2-24\phi_2-22}$	$\frac{5\phi_1+11\phi_2-11}{-3\phi_1^2+8\phi_1\phi_2+19\phi_1+41\phi_2^2-24\phi_2-22}$	$-\frac{3\phi_2-2\phi_1+2}{-3\phi_1^2+8\phi_1\phi_2+19\phi_1+41\phi_2^2-24\phi_2-22}$
z_4	$-\frac{3\phi_1^2+8\phi_1\phi_2-\phi_1+5\phi_2^2+5\phi_2-11}{-6\phi_1^2+16\phi_1\phi_2+38\phi_1+82\phi_2^2-48\phi_2-44}$	$\frac{5\phi_1+11\phi_2-11}{-3\phi_1^2+8\phi_1\phi_2+19\phi_1+41\phi_2^2-24\phi_2-22}$	$-\frac{4\phi_1+10\phi_2-11}{-3\phi_1^2+8\phi_1\phi_2+19\phi_1+41\phi_2^2-24\phi_2-22}$	$-\frac{\phi_1+\phi_2}{-3\phi_1^2+8\phi_1\phi_2+19\phi_1+41\phi_2^2-24\phi_2-22}$
z_5	$-\frac{9\phi_1-13\phi_2+\phi_1\phi_2+21\phi_2^2-12}{2(-3\phi_1^2+8\phi_1\phi_2+19\phi_1+41\phi_2^2-24\phi_2-22)}$	$-\frac{3\phi_2-2\phi_1+2}{-3\phi_1^2+8\phi_1\phi_2+19\phi_1+41\phi_2^2-24\phi_2-22}$	$-\frac{\phi_1+\phi_2}{-3\phi_1^2+8\phi_1\phi_2+19\phi_1+41\phi_2^2-24\phi_2-22}$	$\frac{4\phi_2-\phi_1+2}{-3\phi_1^2+8\phi_1\phi_2+19\phi_1+41\phi_2^2-24\phi_2-22}$

	z_1	RHS
w_1	$\frac{4(32\phi_2-7\phi_1+29)}{-3\phi_1^2+8\phi_1\phi_2+19\phi_1+41\phi_2^2-24\phi_2-22}$	$-\frac{16\phi_2-68\phi_1+84\rho-148\phi_1\phi_2-111\phi_1\rho+192\phi_2\rho+75\phi_1\phi_2^2-18\phi_1^2\phi_2+9\phi_1^2\rho-225\phi_2^2\rho+49\phi_1^2-3\phi_1^3-96\phi_2^2+54\phi_1\phi_2\rho+76}{2(-3\phi_1^2+8\phi_1\phi_2+19\phi_1+41\phi_2^2-24\phi_2-22)}$
z_2	$\frac{3\phi_1^2+18\phi_1\phi_2-37\phi_1-75\phi_2^2+64\phi_2+28}{2(-3\phi_1^2+8\phi_1\phi_2+19\phi_1+41\phi_2^2-24\phi_2-22)}$	$-\frac{(\phi_1-3\rho+2)(9\phi_1^3-9\phi_1^2\phi_2-33\phi_1^2-87\phi_1\phi_2^2+21\phi_1\phi_2+22\phi_1-59\phi_2^3+13\phi_2^2+50\phi_2+5)}{4(-3\phi_1^2+8\phi_1\phi_2+19\phi_1+41\phi_2^2-24\phi_2-22)}$
z_3	$-\frac{2(\phi_1-20\phi_2+5)}{-3\phi_1^2+8\phi_1\phi_2+19\phi_1+41\phi_2^2-24\phi_2-22}$	$\frac{(\phi_1-3\rho+2)(-6\phi_1^2-\phi_1\phi_2+11\phi_1+15\phi_2^2-16\phi_2+1)}{2(-3\phi_1^2+8\phi_1\phi_2+19\phi_1+41\phi_2^2-24\phi_2-22)}$
z_4	$-\frac{2\phi_1+14\phi_2-22}{-3\phi_1^2+8\phi_1\phi_2+19\phi_1+41\phi_2^2-24\phi_2-22}$	$\frac{(\phi_1-3\rho+2)(3\phi_1^2+8\phi_1\phi_2-\phi_1+5\phi_2^2+5\phi_2-11)}{2(-3\phi_1^2+8\phi_1\phi_2+19\phi_1+41\phi_2^2-24\phi_2-22)}$
z_5	$-\frac{2(13\phi_2-2\phi_1+6)}{-3\phi_1^2+8\phi_1\phi_2+19\phi_1+41\phi_2^2-24\phi_2-22}$	$\frac{(\phi_1-3\rho+2)(9\phi_1-13\phi_2+\phi_1\phi_2+21\phi_2^2-12)}{2(-3\phi_1^2+8\phi_1\phi_2+19\phi_1+41\phi_2^2-24\phi_2-22)}$

Table A.8 $T^{ph1}_{\mathcal{B}^{2.1}_{vii}}(\phi, \rho)$ (Due to space limitations, we only display columns associated with nonbasic variables.)

	w_2	w_3	w_5	z_1	z_4	RHS
w_1	$\frac{2\phi_1+20\phi_2-25}{2(4\phi_1+10\phi_2-11)}$	$\frac{6}{4\phi_1+10\phi_2-11}$	$-\frac{6}{4\phi_1+10\phi_2-11}$	$\frac{36}{4\phi_1+10\phi_2-11}$	$-\frac{2(\phi_1+7\phi_2-11)}{4\phi_1+10\phi_2-11}$	$\frac{\phi_1}{2}-\frac{3\rho}{2}-\frac{3(\phi_1+5\phi_2-6)(\phi_1-3\rho+2)}{4\phi_1+10\phi_2-11}+4$
z_2	$\frac{\phi_2}{4}-\frac{\phi_1}{2}-\frac{\left(\frac{\phi_1}{2}+\frac{5\phi_2}{2}-3\right)^2}{4\phi_1+10\phi_2-11}-1$	$\frac{3\phi_1+5\phi_2-5}{2(4\phi_1+10\phi_2-11)}$	$\frac{\phi_1+5\phi_2-6}{8\phi_1+20\phi_2-22}$	$-\frac{2\phi_1+20\phi_2-25}{2(4\phi_1+10\phi_2-11)}$	$\frac{3\phi_1^2+8\phi_1\phi_2-\phi_1+5\phi_2^2+5\phi_2-11}{2(4\phi_1+10\phi_2-11)}$	$-\frac{(\phi_1-3\rho+2)(-9\phi_1^2-26\phi_1\phi_2+18\phi_1-15\phi_2^2+9\phi_2+8)}{4(4\phi_1+10\phi_2-11)}$
z_3	$-\frac{3\phi_1+5\phi_2-5}{2(4\phi_1+10\phi_2-11)}$	$\frac{1}{4\phi_1+10\phi_2-11}$	$-\frac{1}{4\phi_1+10\phi_2-11}$	$\frac{6}{4\phi_1+10\phi_2-11}$	$\frac{\phi_1+\phi_2}{4\phi_1+10\phi_2-11}+1$	$\frac{(3\phi_1+5\phi_2-5)(\phi_1-3\rho+2)}{2(4\phi_1+10\phi_2-11)}$
w_4	$\frac{3\phi_1^2+8\phi_1\phi_2-\phi_1+5\phi_2^2+5\phi_2-11}{2(4\phi_1+10\phi_2-11)}$	$-\frac{\phi_1+\phi_2}{4\phi_1+10\phi_2-11}-1$	$\frac{\phi_1+\phi_2}{4\phi_1+10\phi_2-11}$	$\frac{2(\phi_1+7\phi_2-11)}{4\phi_1+10\phi_2-11}$	$\phi_1-4\phi_2-\frac{(\phi_1+\phi_2)^2}{4\phi_1+10\phi_2-11}-2$	$\frac{(\phi_1+\phi_2)(\phi_1+5\phi_2-6)(\phi_1-3\rho+2)}{2(4\phi_1+10\phi_2-11)}-\frac{(\phi_1+\phi_2+1)(\phi_1-3\rho+2)}{2}$
z_5	$-\frac{\phi_1+5\phi_2-6}{8\phi_1+20\phi_2-22}$	$-\frac{1}{4\phi_1+10\phi_2-11}$	$\frac{1}{4\phi_1+10\phi_2-11}$	$-\frac{6}{4\phi_1+10\phi_2-11}$	$-\frac{\phi_1+\phi_2}{4\phi_1+10\phi_2-11}$	$\frac{(\phi_1+5\phi_2-6)(\phi_1-3\rho+2)}{2(4\phi_1+10\phi_2-11)}$

Table A.9 $T^{ph1}_{\mathcal{B}^{2.1}_{viii}}(\phi, \rho)$

	w_1	w_2	w_3	w_4	w_5	z_1	z_2	z_3	z_4	z_5	RHS
w_1	1	$\frac{5}{2}$	0	0	0	0	0	-6	-8	0	$\frac{15\rho}{2} - \frac{5\phi_1}{2} - 2$
z_2	0	$\frac{\phi_2}{4} - \frac{3}{4}$	0	0	$\frac{1}{2}$	$-\frac{5}{2}$	1	$\frac{5}{2} - \frac{5\phi_2}{2} - \frac{3\phi_1}{2}$	$3 - \frac{5\phi_2}{2} - \frac{3\phi_1}{2}$	0	$-\frac{(\phi_2-3)\left(\frac{\phi_1}{2}-\frac{3\rho}{2}+1\right)}{2}$
w_3	0	$\frac{5}{2} - \frac{5\phi_2}{2} - \frac{3\phi_1}{2}$	1	0	-1	6	0	$4\phi_1 + 10\phi_2 - 11$	$5\phi_1 + 11\phi_2 - 11$	0	$\frac{(3\phi_1+5\phi_2-5)(\phi_1-3\rho+2)}{2}$
w_4	0	$3 - \frac{5\phi_2}{2} - \frac{3\phi_1}{2}$	0	1	-1	8	0	$5\phi_1 + 11\phi_2 - 11$	$7\phi_1 + 8\phi_2 - 13$	0	$\frac{(3\phi_1+5\phi_2-6)(\phi_1-3\rho+2)}{2}$
z_5	0	$-\frac{1}{2}$	0	0	0	0	0	1	1	1	$\frac{\phi_1}{2} - \frac{3\rho}{2} + 1$

Table A.10 $T_{\mathcal{B}_1^{2.1}}(\phi)$ (Due to space limitations, we only display columns associated with nonbasic variables.)

	w_1	w_2	w_3	w_4	w_5
z_1	$\frac{-3\phi_1^2+8\phi_1\phi_2+19\phi_1+41\phi_2^2-24\phi_2-22}{4(32\phi_2-7\phi_1+29)}$	$-\frac{3\phi_1^2+18\phi_1\phi_2-37\phi_1-75\phi_2^2+64\phi_2+28}{8(32\phi_2-7\phi_1+29)}$	$-\frac{\phi_1-20\phi_2+5}{2(32\phi_2-7\phi_1+29)}$	$-\frac{\phi_1+7\phi_2-11}{64\phi_2-14\phi_1+58}$	$-\frac{13\phi_2-2\phi_1+6}{2(32\phi_2-7\phi_1+29)}$
z_2	$-\frac{3\phi_1^2+18\phi_1\phi_2-37\phi_1-75\phi_2^2+64\phi_2+28}{8(32\phi_2-7\phi_1+29)}$	$-\frac{-81\phi_1^2+288\phi_1\phi_2+69\phi_1+47\phi_2^2+388\phi_2+62}{16(32\phi_2-7\phi_1+29)}$	$-\frac{29\phi_1-60\phi_2+9}{4(32\phi_2-7\phi_1+29)}$	$\frac{13\phi_1-5\phi_2+43}{4(32\phi_2-7\phi_1+29)}$	$\frac{2\phi_1+9\phi_2+24}{4(32\phi_2-7\phi_1+29)}$
z_3	$\frac{\phi_1-20\phi_2+5}{2(32\phi_2-7\phi_1+29)}$	$\frac{29\phi_1-60\phi_2+9}{4(32\phi_2-7\phi_1+29)}$	$-\frac{16}{32\phi_2-7\phi_1+29}$	$\frac{12}{32\phi_2-7\phi_1+29}$	$\frac{4}{32\phi_2-7\phi_1+29}$
z_4	$\frac{\phi_1+7\phi_2-11}{64\phi_2-14\phi_1+58}$	$-\frac{13\phi_1-5\phi_2+43}{4(32\phi_2-7\phi_1+29)}$	$\frac{12}{32\phi_2-7\phi_1+29}$	$-\frac{9}{32\phi_2-7\phi_1+29}$	$-\frac{3}{32\phi_2-7\phi_1+29}$
z_5	$\frac{13\phi_2-2\phi_1+6}{2(32\phi_2-7\phi_1+29)}$	$-\frac{2\phi_1+9\phi_2+24}{4(32\phi_2-7\phi_1+29)}$	$\frac{4}{32\phi_2-7\phi_1+29}$	$-\frac{3}{32\phi_2-7\phi_1+29}$	$-\frac{1}{32\phi_2-7\phi_1+29}$

	RHS
z_1	$\frac{3\phi_1^3+18\phi_1^2\phi_2-49\phi_1^2-75\phi_1\phi_2^2+148\phi_1\phi_2+68\phi_1+96\phi_2^2-16\phi_2-76}{8(32\phi_2-7\phi_1+29)}$
z_2	$\frac{-81\phi_1^3+288\phi_1^2\phi_2-111\phi_1^2+47\phi_1\phi_2^2+856\phi_1\phi_2+422\phi_1+544\phi_2^2+392\phi_2-44}{16(32\phi_2-7\phi_1+29)}$
z_3	$-\frac{61\phi_1-60\phi_1\phi_2+29\phi_1^2-12}{4(32\phi_2-7\phi_1+29)}$
z_4	$\frac{75\phi_1+32\phi_2-5\phi_1\phi_2+13\phi_1^2+20}{4(32\phi_2-7\phi_1+29)}$
z_5	$\frac{16\phi_1+96\phi_2+9\phi_1\phi_2+2\phi_1^2+84}{4(32\phi_2-7\phi_1+29)}$

Appendix B
Tableaux for Example 2.2

Table B.1 $T_{\mathcal{B}_0^{2.2}}(\phi, \sigma)$

	w_1	w_2	w_3	w_4	z_1	z_2	z_3	z_4	RHS
w_1	1	0	0	0	0	0	2	1	$-\sigma - 1$
w_2	0	1	0	0	0	0	5	$-\phi - 7$	$\phi - \sigma - 1$
w_3	0	0	1	0	-1	-3	0	0	$-18\sigma - 34$
w_4	0	0	0	1	-1	$\phi + 5$	0	0	$-9\sigma - 17$

Table B.2 $T_{\mathcal{B}_1^{2.2}}(\phi, \sigma)$

	w_1	w_2	w_3	w_4	z_1	z_2	z_3	z_4	RHS
w_1	1	$\frac{1}{\phi+7}$	0	0	0	0	$\frac{5}{\phi+7} + 2$	0	$-\frac{8\sigma+\phi\sigma+8}{\phi+7}$
z_2	0	0	0	$\frac{1}{\phi+5}$	$-\frac{1}{\phi+5}$	1	0	0	$-\frac{9\sigma+17}{\phi+5}$
w_3	0	0	1	$\frac{3}{\phi+5}$	$-\frac{3}{\phi+5} - 1$	0	0	0	$-\frac{(2\phi+13)(9\sigma+17)}{\phi+5}$
z_4	0	$-\frac{1}{\phi+7}$	0	0	0	0	$-\frac{5}{\phi+7}$	1	$\frac{\sigma-\phi+1}{\phi+7}$

Table B.3 $T_{\mathcal{B}_2^{2.2}}(\phi, \sigma)$

	w_1	w_2	w_3	w_4	z_1	z_2	z_3	z_4	RHS
w_1	1	$-\frac{2}{5}$	0	0	0	0	0	$\frac{2\phi}{5} + \frac{19}{5}$	$-\frac{2\phi}{5} - \frac{3\sigma}{5} - \frac{3}{5}$
z_2	0	0	$-\frac{1}{3}$	0	$\frac{1}{3}$	1	0	0	$6\sigma + \frac{34}{3}$
z_3	0	$\frac{1}{5}$	0	0	0	0	1	$-\frac{\phi}{5} - \frac{7}{5}$	$\frac{\phi}{5} - \frac{\sigma}{5} - \frac{1}{5}$
w_4	0	0	$\frac{\phi}{3} + \frac{5}{3}$	1	$-\frac{\phi}{3} - \frac{8}{3}$	0	0	0	$-\frac{(2\phi+13)(9\sigma+17)}{3}$

N. Adelgren, *Advancing Parametric Optimization*, SpringerBriefs in Optimization,
https://doi.org/10.1007/978-3-030-61821-6

Table B.4 $T_{\mathcal{B}_3^{2.2}}(\phi, \sigma)$

	w_1	w_2	w_3	w_4	z_1	z_2	z_3	z_4	RHS
z_1	0	0	$-\frac{\phi+5}{\phi+8}$	$-\frac{3}{\phi+8}$	1	0	0	0	$\frac{(2\phi+13)(9\sigma+17)}{\phi+8}$
z_2	0	0	$-\frac{1}{\phi+8}$	$\frac{1}{\phi+8}$	0	1	0	0	$\frac{9\sigma+17}{\phi+8}$
z_3	$\frac{\phi+7}{2\phi+19}$	$\frac{1}{2\phi+19}$	0	0	0	0	1	0	$-\frac{8\sigma+\phi\sigma+8}{2\phi+19}$
z_4	$\frac{5}{2\phi+19}$	$-\frac{2}{2\phi+19}$	0	0	0	0	0	1	$-\frac{2\phi+3\sigma+3}{2\phi+19}$

Table B.5 $T_{\mathcal{B}_4^{2.2}}(\phi, \sigma)$

	w_1	w_2	w_3	w_4	z_1	z_2	z_3	z_4	RHS
z_1	0	0	0	-1	1	$-\phi-5$	0	0	$9\sigma+17$
w_2	$\phi+7$	1	0	0	0	0	$2\phi+19$	0	$-8\sigma-\phi\sigma-8$
w_3	0	0	1	-1	0	$-\phi-8$	0	0	$-9\sigma-17$
z_4	1	0	0	0	0	0	2	1	$-\sigma-1$

Table B.6 $T_{\mathcal{B}_5^{2.2}}(\phi, \sigma)$

	w_1	w_2	w_3	w_4	z_1	z_2	z_3	z_4	RHS
z_1	0	0	-1	0	1	3	0	0	$18\sigma+34$
w_2	$-\frac{5}{2}$	1	0	0	0	0	0	$-\phi-\frac{19}{2}$	$\phi+\frac{3\sigma}{2}+\frac{3}{2}$
z_3	$\frac{1}{2}$	0	0	0	0	0	1	$\frac{1}{2}$	$-\frac{\sigma}{2}-\frac{1}{2}$
w_4	0	0	-1	1	0	$\phi+8$	0	0	$9\sigma+17$

References

1. Adelgren N (2016) Solution techniques for classes of biobjective and parametric programs. PhD thesis, Clemson University
2. Adelgren N, Wiecek MM (2016) A two-phase algorithm for the multiparametric linear complementarity problem. Eur J Oper Res 254(3):715–738
3. Bank B, Guddat J, Klatte D, Kummer B, Tammer K (1982) Non-linear parametric optimization. Springer, New York
4. Barnett S (1968) A simple class of parametric linear programming problems. Oper Res 16(6):1160–1165
5. Basu S, Pollack R, Roy MF (2006) Algorithms in real algebraic geometry, algorithms and computation in mathematics, vol 10, 2nd edn. Springer, Berlin/Heidelberg. https://doi.org/10.1007/3-540-33099-2
6. Blanchini F (1999) Set invariance in control. Automatica 35(11):1747–1767
7. Bôcher M, Duval EPR (1907) Introduction to higher algebra. Macmillan, New York
8. Bochnak J, Coste M, Roy MF (2013) Real algebraic geometry, vol 36. Springer Science & Business Media, Berlin.
9. Chakraborty B, Nanda S, Biswal M (2004) On the solution of parametric linear complementarity problems. Int J Pure Appl Math 17:9–18
10. Charitopoulos VM, Papageorgiou LG, Dua V (2017) Multi-parametric linear programming under global uncertainty. AIChE J 63(9):3871–3895
11. Columbano S, Fukuda K, Jones CN (2009) An output-sensitive algorithm for multi-parametric LCPs with sufficient matrices. In: Polyhedral computation, vol 48. American Mathematical Society, Providence, pp 73–102
12. Cottle R, Pang JS, Venkateswaran V (1989) Sufficient matrices and the linear complementarity problem. Linear Algebra Appl 114:231–249
13. Cottle RW (2010) A field guide to the matrix classes found in the literature of the linear complementarity problem. J Global Optim 46(4):571–580
14. Cottle RW, Pang JS, Stone RE (2009) The linear complementarity problem. SIAM, Philadelphia
15. Courtillot M (1962) On varying all the parameters in a linear-programming problem and sequential solution of a linear-programming problem. Oper Res 10(4):471–475
16. Dent W, Jagannathan R, Rao M (1973) Parametric linear programming: some special cases. Nav Res Logist Q 20(4):725–728
17. Ehrgott M (2005) Multicriteria optimization, vol 2. Springer, Heidelberg

N. Adelgren, *Advancing Parametric Optimization*, SpringerBriefs in Optimization,
https://doi.org/10.1007/978-3-030-61821-6

18. Filar J, Avrachenkov K, Altman E (1999) An asymptotic simplex method for parametric linear programming. In: Information, decision and control, 1999. IDC 99. Proceedings. 1999. IEEE, pp 427–432
19. Finkelstein B, Gumenok L (1977) Algorithm for solving a linear parametric program when the A-matrix depends upon a parameter. Ekonomicko Matematiceskie Metody 13:342–347
20. Fotiou IA, Rostalski P, Parrilo PA, Morari M (2006) Parametric optimization and optimal control using algebraic geometry methods. Int J Control 79(11):1340–1358
21. Gailly B, Installe M, Smeers Y (2001) A new resolution method for the parametric linear complementarity problem. Eur J Oper Res 128(3):639–646
22. Gerhard J, Jeffrey D, Moroz G (2010) A package for solving parametric polynomial systems. ACM Commun Comput Algebra 43(3/4):61–72
23. Ghaffari-Hadigheh A, Romanko O, Terlaky T (2007) Sensitivity analysis in convex quadratic optimization: simultaneous perturbation of the objective and right-hand-side vectors. Algorithmic Oper Res 2(2):94
24. Ghaffari-Hadigheh A, Romanko O, Terlaky T (2010) Bi-parametric convex quadratic optimization. Optim Methods Softw 25(2):229–245
25. Goh C, Yang X (1996) Analytic efficient solution set for multi-criteria quadratic programs. Eur J Oper Res 92(1):166–181
26. Güler O (2010) Foundations of optimization, vol 258. Springer, New York/Heidelberg
27. den Hertog D, Roos C, Terlaky T (1993) The linear complimentarity problem, sufficient matrices, and the criss-cross method. Linear Algebra Appl 187:1–14
28. Hirschberger M, Steuer RE, Utz S, Wimmer M, Qi Y (2013) Computing the nondominated surface in tri-criterion portfolio selection. Oper Res 61(1):169–183
29. Hirschberger M, Qi Y, Steuer RE (2010) Large-scale MV efficient frontier computation via a procedure of parametric quadratic programming. Eur J Oper Res 204(3):581–588
30. Hladík M (2012) Interval linear programming: a survey. In: Mann ZA (ed) Linear programming – new frontiers in theory and applications, chap 2. Nova Science Publishers, New York, pp 85–120
31. Jayasekara PL, Adelgren N, Wiecek MM (2020) On convex multiobjective programs with application to portfolio optimization. J Multi-Criteria Decis Anal 27(3–4):189–202
32. Jonker P, Still G, Twilt F (2001) One-parametric linear-quadratic optimization problems. Ann Oper Res 101(1–4):221–253
33. Khalilpour R, Karimi I (2014) Parametric optimization with uncertainty on the left hand side of linear programs. Comput Chem Eng 60:31–40
34. Kim C (1971) Parameterizing an activity vector in linear programming. Oper Res 19(7):1632–1646
35. Kolev L, Skalna I (2018) Exact solution to a parametric linear programming problem. Numer Algorithms 78(4):1183–1194
36. Kostreva M (1989) Generalization of Murty's direct algorithm to linear and convex quadratic programming. J Optim Theory Appl 62(1):63–76
37. Lazard D, Rouillier F (2007) Solving parametric polynomial systems. J Symb Comput 42(6):636–667
38. Lemke CE (1965) Bimatrix equilibrium points and mathematical programming. Manag Sci 11(7):681–689
39. Li Z, Ierapetritou MG (2010) A method for solving the general parametric linear complementarity problem. Ann Oper Res 181(1):485–501
40. Maier G (1970) A matrix structural theory of piecewise linear elastoplasticity with interacting yield planes. Meccanica 5(1):54–66
41. MATLAB (2016) version 9.0.0 (R2016a). The MathWorks Inc., Natick. http://citebay.com/how-to-cite/matlab/
42. Murty KG, Yu FT (1997) Linear complementarity, linear and nonlinear programming (Internet Edition)

43. Pistikopoulos EN, Dominguez L, Panos C, Kouramas K, Chinchuluun A (2012) Theoretical and algorithmic advances in multi-parametric programming and control. Comput Manag Sci 9(2):183–203
44. Ritter DMK (1962) Ein verfahren zur lösung parameterabhängiger, nichtlinearer maximum-probleme. Unternehmensforschung 6(4):149–166
45. Rotman JJ (2000) A first course in abstract algebra, 2nd edn. Prentice Hall, Upper Saddle River, NJ.
46. Song Y (2015) Optimization theory and dynamical systems: invariant sets and invariance preserving discretization methods. PhD thesis, Lehigh University
47. Steuer RE, Qi Y, Hirschberger M (2011) Comparative issues in large-scale mean–variance efficient frontier computation. Decis Support Syst 51(2):250–255
48. Tammer K (1998) Parametric linear complementarity problems (chapter). In Mathematical programming with data perturbations. CRC Press, New York, pp 399–418
49. Väliaho H (1979) A procedure for one-parametric linear programming. BIT Numer Math 19(2):256–269
50. Väliaho H (1985) A unified approach to one-parametric general quadratic programming. Math Program 33(3):318–338
51. Väliaho H (1994) A procedure for the one-parametric linear complementarity problem. Optimization 29(3):235–256
52. Väliaho H (1996) Criteria for sufficient matrices. Linear Algebra Appl 233:109–129
53. Väliaho H (1997) Determining the handicap of a sufficient matrix. Linear Algebra Appl 253(1):279–298
54. Vaught R (1974) Invariant sets in topology and logic. Fundamenta Mathematicae 82(3):269–294
55. Whitney H (1943) The general type of singularity of a set of $2n - 1$ smooth functions of n variables. Duke Math J 10(1):161–172
56. Willner LB (1967) On parametric linear programming. SIAM J Appl Math 15(5):1253–1257
57. Xiao B (1995) The linear complementarity-problem with a parametric input. Eur J Oper Res 81(2):420–429